HISTOSOLS: Their Characteristics, Classification, and Use

HISTOSOLS

Their Characteristics, Classification, and Use

Proceedings of a symposium held on 30 October 1972 during the annual meeting of the Soil Science Society of America in Miami Beach, Florida. Sponsored by Division S-5, Soil Genesis, Morphology, and Classification.

Editorial Committee

A. R. AANDAHL, chairman

Professor of Agronomy, University of Nebraska, Lincoln, Nebraska

S. W. BUOL

Professor of Soil Genesis and Classification, North Carolina State University, Raleigh, North Carolina

D. E. HILL

Associate Soil Scientist, The Connecticut Agricultural Experiment Station, New Haven, Connecticut

H. H. BAILEY

Professor of Soil Classification, University of Kentucky, Lexington, Kentucky

Coordinating Editor

MATTHIAS STELLY

Managing Editor

RICHARD C. DINAUER

Number 6 in the

SSSA SPECIAL PUBLICATION SERIES

Soil Science Society of America, Inc., publisher

Madison, Wisconsin USA

1974

The cover design is by Ruth C. Poulsen, staff artist with the American Society of Agronomy, Madison, Wis. The composition for the text pages was done by Mrs. Linda Nelson on an IBM Selectric Composer using the Baskerville type font.

Soil Science Society of America, Inc.
677 South Segoe Road, Madison, Wisconsin 53711 USA

Library of Congress Catalog Card Number: 74-81344

Printed in the United States of America

CONTENTS

6 Physical, Chemical, Elemental, and Oxygen-Containing Functional Group Analysis of Selected Florida Histosols

L. W. ZELAZNY and V. W. CARLISLE

7 Some Engineering Aspects of Peat Soils

I. C. MAC FARLANE and G. P. WILLIAMS

FOREWORD

Man's effort to understand the world around him requires the ordering of things into comprehensible units. Although pedologists have been evolving a classification for soils for many years, a comprehensive system for organic soils (Histosols) was largely neglected until the 1950's. During the past 20 years, an acceptable classification system for Histosols has been developed.

It is important that soil scientists be familiar with this new system because Histosols are important soils both in extent and use. They are favored soils for many special uses, and they produce a significant part of the world's food and forest products. Sustained use of these fragile soils requires that their characteristics be understood and classified, and that effective management systems be developed for the various categories of Histosols.

A symposium held as a part of the program of the annual meeting of the Soil Science Society of America in October 1972 provided a forum for presentation of many excellent papers on the characteristics, classification, uses, and analysis procedures for Histosols. The papers from this symposium comprise this special publication.

The Soil Science Society of America is pleased to sponsor this publication with the belief that it will help refine and improve soil classification and increase understanding, communication, and wise use of Histosols.

The Society is indebted to the organizers, the authors, the editorial committee, and the headquarters staff for their diligent and dedicated efforts which made this publication possible.

Lubbock, Texas
May 1974

ANSON R. BERTRAND, *president*
Soil Science Society of America

PREFACE

Organic soils (Histosols) have long been recognized as having properties distinctly different from those of mineral soils. Highway engineers tried to avoid them or to replace them with mineral soils or materials, while truck farmers preferred them to mineral soils for many crops.

The classification of organic soils, however, was neglected until about two decades ago. Generally, most of the distinctions were limited to a recognition of acid or calcareous peats and mucks. The acreages of organic soils are small throughout most of the lands used for growing cultivated crops such as corn, wheat, and cotton. The soils used for growing cultivated crops were the main ones that soil scientists studied and classified.

When the present system of soil classification was developed during the 1950's and 1960's, soil scientists realized that they must concentrate much of their effort on organic soils if they were to have a classification of organic soils that was comparable to their classification of mineral soils.

Consequently, during the 1950's and 1960's the efforts of many soil scientists were directed toward the study of organic soils. They searched for characteristics that could be recognized in the field with or without simple tests and that could be studied in more detail in the laboratory. Using these characteristics they developed different kinds of taxa and organized them into possible systems of classification. These systems were then tested for their usefulness in making predictions of the behavior of organic soils which would best serve as guides to plan the use and management of these soils.

When the program for the 1972 annual meeting of the Soil Science Society of America was being prepared, it was obvious that the progress made by these students of organic soils should be made available to other soil scientists and people interested in organic soils. Therefore, a symposium on organic soils was organized by Harry Hudson Bailey of the University of Kentucky and was included in the programs for Division S-5 of SSSA under the direction of division program chairman, Robert B. Grossman, SCS-USDA, Lincoln, Nebraska. The excellent papers of this symposium provided the material for this publication.

Lincoln, Nebraska
April 1974

ANDREW R. AANDAHL
Chairman, Editorial Committee

CONVERSION FACTORS FOR ENGLISH AND METRIC UNITS

To convert column 1 into column 2, multiply by	Column 1	Column 2	To convert column 2 into column 1, multiply by
	Length		
0.621	kilometer, km	mile, mi	1.609
1.094	meter, m	yard, yd	0.914
0.394	centimeter, cm	inch, in	2.54
	Area		
0.386	kilometer^2, km^2	mile^2, mi^2	2.590
247.1	kilometer^2, km^2	acre, acre	0.00405
2.471	hectare, ha	acre, acre	0.405
	Volume		
0.00973	meter^3, m^3	acre-inch	102.8
3.532	hectoliter, hl	cubic foot, ft^3	0.2832
2.838	hectoliter, hl	bushel, bu	0.352
0.0284	liter	bushel, bu	35.24
1.057	liter	quart (liquid), qt	0.946
	Mass		
1.102	ton (metric)	ton (English)	0.9072
2.205	quintal, q	hundredweight, cwt (short)	0.454
2.205	kilogram, kg	pound, lb	0.454
0.035	gram, g	ounce (avdp), oz	28.35
	Pressure		
14.50	bar	lb/inch^2, psi	0.06895
0.9869	bar	atmosphere,* atm	1.013
0.9678	kg(weight)/cm^2	atmosphere,* atm	1.033
14.22	kg(weight)/cm^2	lb/inch^2, psi	0.07031
14.70	atmosphere,* atm	lb/inch^2, psi	0.06805
	Yield or Rate		
0.446	ton (metric)/hectare	ton (English)/acre	2.240
0.892	kg/ha	lb/acre	1.12
0.892	quintal/hectare	hundredweight/acre	1.12
1.15	hectoliter/ha, hl/ha	bu/acre	0.87
	Temperature		
$\left(\frac{9}{5}\ ^\circ C\right) + 32$	Celsius	Fahrenheit	$\frac{5}{9}\ (^\circ F - 32)$
	−17.8C	0F	
	0C	32F	
	20C	68F	
	100C	212F	
	Water Measurement		
8.108	hectare-meters, ha-m	acre-feet	0.1233
97.29	hectare-meters, ha-m	acre-inches	0.01028
0.08108	hectare-centimeters, ha-cm	acre-feet	12.33
0.973	hectare-centimeters, ha-cm	acre-inches	1.028
0.00973	meters^3, m^3	acre-inches	102.8
0.981	hectare-centimeters/hour, ha-cm/hour	feet^3/sec	1.0194
440.3	hectare-centimeters/hour, ha-cm/hour	U.S. gallons/min	0.00227
0.00981	meters^3/hour, m^3/hour	feet^3/sec	101.94
4.403	meters^3/hour, m^3/hour	U.S. gallons/min	0.227

* The size of an "atmosphere" may be specified in either metric or English units.

Criteria Used in Soil Taxonomy to Classify Organic Soils[1]

1

WILLIAM E. MC KINZIE[2]

ABSTRACT

A new system has been developed for classifying organic soils (Histosols) based on quantitative criteria that can be determined by visual observations or by simple tests. Histosols are dominantly organic material and commonly called bogs, coastal marshes, moors, muskegs, peats, or mucks. The organic material exceeds 12 to 18% organic carbon by weight depending on the clay content of the mineral fraction and constitutes more than 50% of the upper 80 cm unless the depth to rock or to fragmental materials is less than 80 cm. Three kinds of organic soil materials (fibric, hemic and sapric) based on the degree of decomposition of the plant materials are used in the classification of the higher categories. Limnic (sedimentary) materials consisting of marl, diatomaceous earth and coprogenous earth are also used in the lower categories. Particle size, mineralogy, reaction, temperature, and soil depth are recognized at the family level.

INTRODUCTION

There have been numerous criteria used in classifying organic soils. The majority of these classification systems were based on a common criteria or a combination of criteria and were developed for a specific region or purpose.

Farnham and Finney (1965) in their publication "Classification and Properties of Organic Soil" have made an extensive review of the literature in a search for the various criteria used in the many systems proposed for classifying organic soils. A summary of the previous systems would include

1) Topographical features (Shaler, 1890; Weber, 1903);
2) Surface vegetation features (Ogg, 1939; Radforth, 1952-1953; Heinselman, 1963);
3) Chemical properties (Sukachev, 1926; Alway, 1920; Harmer, 1941; Hygard, 1954; Godwin, 1941);
4) Botanical origin (Davis, 1946; Rigg, 1958; Davis & Lucas, 1959);
5) Morphology (Post, 1926; Veatch, 1953; Dachnowski, 1924; Troels-Smith, 1955); and
6) Gentic processes (Veatch, 1927; Dachnowski, 1940; Kazakov, 1953; Kubiena, 1953; Fraser, 1943-1954).

[1] Contribution from Soil Conservation Service, USDA.

[2] Soil Correlator, Soil Conservation Service, Midwest Technical Service, Center, Lincoln, Nebraska.

A review of these former systems indicated that there was a real need for a new system based on quantitative criteria and on criteria that could be determined in the field by visual observations or by simple field tests. The Farnham and Finney publication was used as a base for developing a new system. The system that appears in *Soil Taxonomy* (Soil Survey Staff, 1974) is the result of the modification and changes that were made by the efforts of many workers interested in the classification of organic soils. This paper outlines the criteria used in *Soil Taxonomy* (Soil Survey Staff, 1974) in the classification of Histosols.

CLASSIFICATION OF HISTOSOLS

Histosols are soils that are dominantly organic and commonly called bogs, coastal marshes, moors, muskegs, peats, or mucks. A few consist of shallow organic materials resting on rock or rubble. The organic materials that constitute Histosols contain at least from 12 to 18% organic carbon, by weight, depending on the clay content of the mineral fraction (Figure 1). Histosols are readily distinguished from mineral soils by bulk density. It is a general rule that a soil is classed as an organic soil (Histosol) either if more than 50%

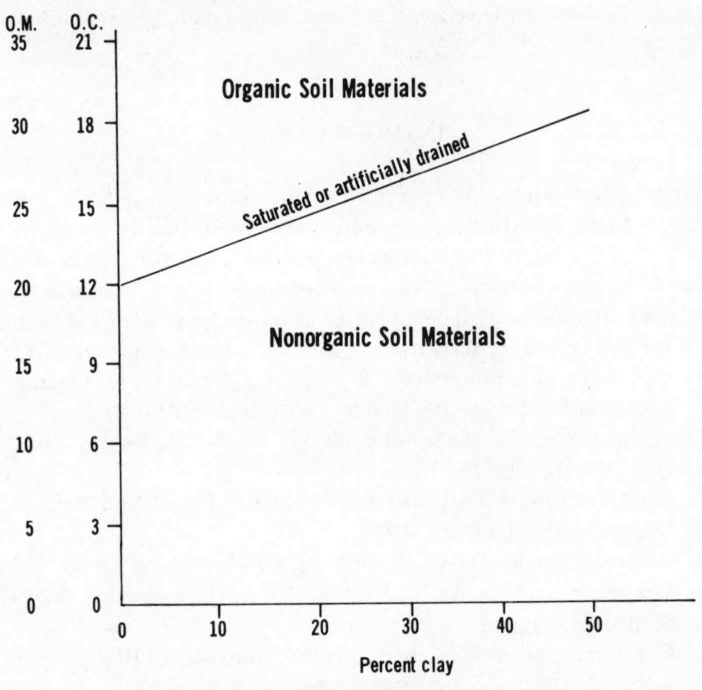

Figure 1. Organic carbon requirements for organic soil materials.

of the upper 80 cm of soil is organic material or if organic material of any thickness rests on rock or in fragmental material having interstices filled with organic materials (Figure 2).

In addition to the required organic carbon content, Histosols are organic soils that, except for thin mineral layers, extend from the surface to one of the following (Figure 3):

1) A depth of 60 cm or more if 75% or more of the volume is fibric sphagnum or moss or if the bulk density is less than $0.1 g/c^3$;
2) A depth of 40 cm if the organic soil material is saturated with water for long periods (6 months) or is artificially drained, and the organic material has a bulk density of 0.1 or more;
3) A depth of 10 cm or less above a lithic or paralithic contact, provided the thickness of the organic soil material is more than twice that of the mineral material above the contact;
4) Any depth if the organic material rests on fragmental material (gravel, stones, or cobbles) in which the interstices are filled or partly filled with organic materials or rests on a lithic or paralithic contact.

ORGANIC SOIL MATERIALS

Three basic kinds of organic soil materials are distinguished—fibric, hemic, and sapric according to the degree of decomposition of the original plant ma-

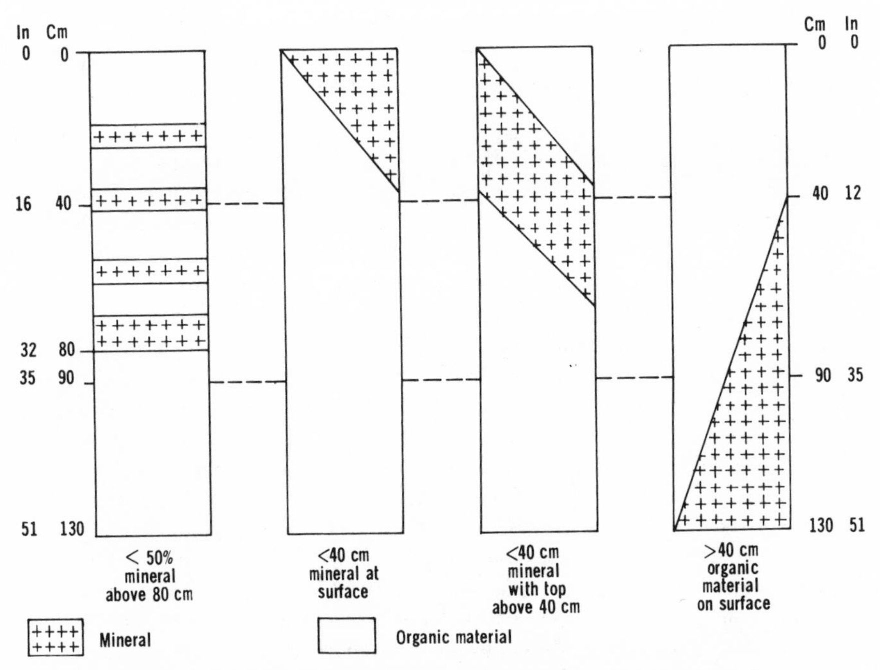

Figure 2. Distribution of mineral and organic layers in organic soils.

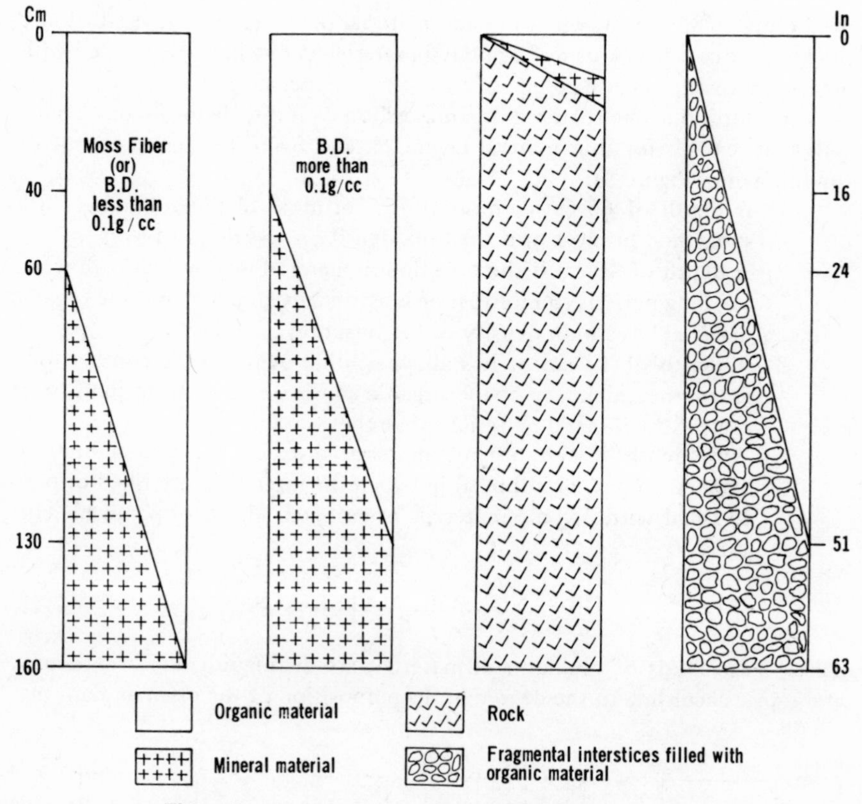

Figure 3. Thickness requirements of organic materials.

terials. Fiber content is used in the definition of these materials. A fiber is a fragment or piece of plant tissue, excluding live roots, that is large enough to be retained on a 100-mesh sieve (openings 0.15 mm in diameter) and that retains recognizable cellular structure of the plant from which it came. The material is screened after dispersion in sodium hexametaphosphate. Undecomposed coarse fragments, i.e., greater than 2 cm in cross section or smallest dimension, are excluded. However, larger fragments that can be crushed and shredded with the fingers are included in fibers.

The degree of decomposition of organic materials is indicated by the content of fibers. If material is highly decomposed, fibers are nearly absent. If it is only slightly decomposed, more of the volume, exclusive of the coarse fragments, normally consists of fibers. If the organic materials are moderately decomposed, the fibers may be largely preserved but they are easily broken down by rubbing. For this reason, the percentage of fibers that do not break down when rubbed gives the most realistic field estimate of the degree of decomposition. In addition, the bulk density and hence the amount of subsidence after drainage is more closely related to the content of fiber after rubbing than to the content of fiber in the undisturbed condition. In the

KIND ORGANIC MATERIAL	FIBER CONTENT		SODIUM PYROPHOSPHATE EXTRACT COLOR (Munsell Notation)
	UNRUBBED (Volume)*	RUBBED (Volume)	
SAPRIC	<1/3	——	——
	>1/3	<1/6	6/3, 5/2, 4/1, 3/1, 2/1, or darker
HEMIC	1/3 to 2/3	>1/6	——
	>2/3	<2/5	——
	——	——	Outside range of fibric or sapric materials
FIBRIC	>2/3	>3/4	——
	>2/3	>2/5 &<3/4	7/1, 7/2, 8/1, 8/2, 8/3

* Not diagnostic in classification of organic materials

LIMNIC MATERIALS

COPROGENOUS EARTH (SEDIMENTARY PEAT), DIATOMACEOUS EARTH, AND MARL.

Figure 4. Organic soil materials.

field the rubbed fiber content is determined by using a small volume of wet material and rubbing it about 10 times, with firm pressure, between the thumb and fingers. The material is then molded into a spherical mass, broken, and examined by use of a hand lens of 10-power or more to estimate the fiber content. In the laboratory, measured amounts are used prior to rubbing and following rubbing and washing. The rubbed materials remaining are also measured and compared to the original volume. The procedure is outlined by W. C. Lynn et al. in chapter 2 of this book. Figure 4 shows the distinguishing characteristics of the three kinds of organic materials.

Fibric soil materials are the least decomposed of all of the organic soil materials. They contain large amounts of fiber which is well-preserved and readily identifiable as to botanical origin. They have low bulk densities and high water contents when saturated. Hemic soil materials are intermediate in degree of decomposition between the less decomposed fibric and the more decomposed sapric materials. Sapric soil materials are the most highly decomposed of the organic materials. They have the least amount of plant fiber, may have the highest bulk density values, and the lowest water contents at saturation of any on a dry weight basis.

LIMNIC MATERIALS

Provisions for the classification of soils formed entirely in limnic materials namely coprogenous earth (sedimentary peat), marl, and diatomaceous earth, still remains to be added; however, soils having formed in part from limnic materials are used to separate classes in the lower categories. Limnic materials include both organic and inorganic materials either (i) deposited in water by precipitation or by action of aquatic organisms such as algae or diatoms, or (ii) derived from underwater and floating aquatic plants subsequently modified by aquatic animals. Limnic materials include marl, diatomaceous earth, coprogenous earth (sedimentary peat), and possibly others. Except for some of the coprogenous earths, most of these limnic materials are inorganic. Diatomite is highly siliceous; and marl is mainly calcium carbonate.

CONTROL SECTION

The Histosol order is defined on the thickness of organic materials over limnic materials, mineral materials, water, or permafrost. A control section has been established for the taxonomy of Histosols. It is either 130 cm or 160 cm thick, depending on the kind of material, provided that no lithic or paralithic contact, thick layer of water, or frozen soil is within those limits.

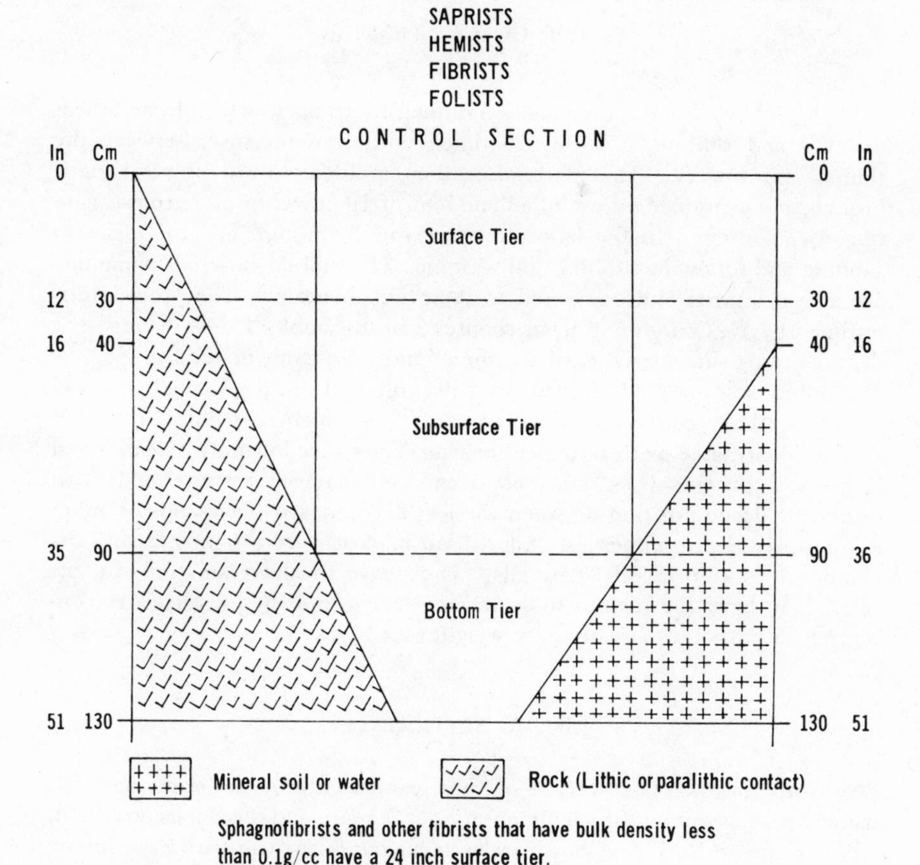

Figure 5. Suborders of Histosols.

The control section has been divided somewhat arbitrarily into three tiers: surface, subsurface, and bottom tiers (Figure 5).

CATEGORIES

Taxa in suborder, great group, and subgroup categories parallel those of the mineral soils. The suborder level is determined by the degree of decomposition of the organic material in the subsurface tier if that tier is organic throughout its thickness. If a mineral layer(s) more than 40 cm thick has an upper boundary in the subsurface tier, the suborder is based on the degree of decomposition of the organic material between the contact of this material and the surface. If a lithic (rock) or paralithic (lithiclike) contact has an upper boundary in the subsurface or surface tier, the suborder is based on degree of decomposition between the contact and the surface (Figure 5).

Suborders

Suborders of Histosols that are saturated with water 6 months or more of the year or have artificial drainage are as follows (Dominant as used in the following context means the most abundant.):

SAPRISTS

Sapric soil materials dominant in the organic part of the control section as outlined above.

HEMISTS

Hemic soil materials dominant in the organic part of the control section as outlined above and/or have a sulfuric horizon that has its upper boundary within 50 cm of the surface or sulfidic materials within 1 m of the surface. A sulfuric horizon has both a pH of less than 3.5 (1:1 in water) and jarosite mottles.

FIBRISTS

Fibric soil materials dominant in the organic part of the control section as outlined above or have fibric material that is 75% or more (by volume) derived from Sphagnum (spp.) and that either makes up the upper 90 cm or more of the soil or rests on a lithic or paralithic contact, fragmental materials or mineral soil materials.

Suborders of Histosols that are never saturated with water for more than a few days following heavy rains.

GREAT GROUP	SOIL TEMPERATURE REGIME	TEMPERATURE CLASSES
CRYOSAPRISTS CRYOHEMISTS CRYOFIBRISTS CRYOFOLISTS SPHAGNOFIBRISTS	MEAN ANNUAL SOIL TEMPERATURE OF LESS THAN 8°C; AND (1) ARE FROZEN IN SOME LAYER WITHIN CONTROL SECTION; OR (2) HAVE AN OCEANIC CLIMATE WITH NO FROST BELOW 5 CM.	FRIGID ISOFRIGID
BOROSAPRISTS BOROHEMISTS BOROFIBRISTS BOROFOLISTS SPHAGNOFIBRISTS	MEAN ANNUAL SOIL TEMPERATURE THAT IS LESS THAN 8°C	FRIGID
MEDISAPRISTS MEDIHEMISTS MEDIFIBRISTS	MEAN ANNUAL SOIL TEMPERATURE THAT IS GREATER THAN 8°C	MESIC THERMIC HYPERTHERMIC
TROPOSAPRISTS TROPOHEMISTS TROPOFIBRISTS TROPOFOLISTS	MEAN ANNUAL SOIL TEMPERATURE OF 8°C OR MORE AND LESS THAN 5°C DIFFERENCE BETWEEN MEAN SUMMER AND WINTER SOIL TEMPERATURES	ISOMESIC ISOTHERMIC ISOHYPERTHERMIC
SULFIHEMISTS SULFOHEMISTS	NO RESTRICTIONS ON CLIMATE	

Figure 6. Great groups of Histosols.

FOLISTS

Consists primarily of O horizons derived from leaf litter, twigs, and branches resting on rock or on fragmental materials that consist of gravel, stones, and boulders in which the interstices are partly filled or filled with organic materials.

Great Groups

The great group level is determined largely by the soil temperature regime (cryo, boro, medi, and tropo). In addition, the presence of sulfidic materials or a sulfuric horizon is used in defining great groups of Hemists and also a surface mantle of fibric Sphagnum (spp.) moss is used in defining great groups of Fibrists (Figure 6).

Subgroups

The subgroup level is determined by a number of properties. Subgroup intergrades to other great groups of organic soils are based on the presence of more than one kind of organic material in the control section, or on the presence of a mineral layer between 5 and 30 cm thick or two or more thin mineral layers (Fluvaquentic). Subgroup extragrades are based on the presence

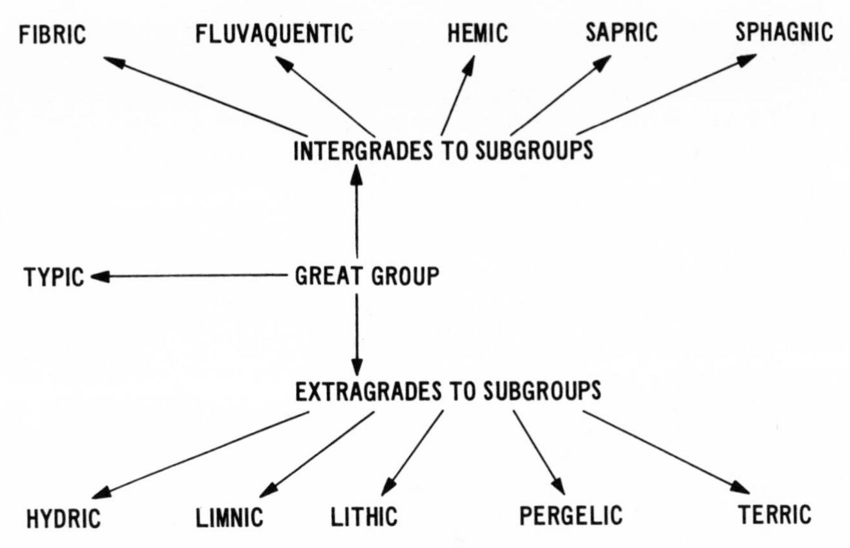

Figure 7. Subgroups of Histosols.

of a mineral (terric) or a limnic layer, the presence of rock (lithic), permafrost (pergelic), and open water. Figure 7 shows the subgroups that have been recognized.

Subgroups are subdivided into families based on the following differentiae:

PARTICLE-SIZE CLASSES

Refers only to the mineral layer of terric subgroups.

Fragmental
Loamy-skeletal or clayey-skeletal
Sandy or sandy-skeletal
Loamy
Clayey

The meaning of each of these terms is the same as defined for particle-size classic of mineral soils. The term refers to the average particle size of the upper 30 cm of the mineral layer or that part of the mineral layer that is within the control section, whichever is thicker.

MINERALOGY CLASSES

Mineralolgy classes of Histosols are of four kinds based on the nature of the subgroup or great group:

Ferrihumic—These contain authigenic deposits (formed in place) of hydrated iron oxides mixed with varying kinds or amounts of organic material.

Terric Subgroups—Mineralogy applied only to the mineral parts of the soil for which a particle-size class has been used.

Limnic Subgroups—Refers to limnic layers within the control section that are 5 cm or more thick.

Coprogenous—Limnic materials consist of coprogenous earth.
Diatomaceous—Limnic materials consist of diatomaceous earth.
Marly—Limnic materials consist of marl.

REACTION CLASSES

Euic—pH is 4.5 or more (in 0.01*M* $CaCl_2$) in at least some part of subsurface tier or the surface and subsurface tiers if a mineral layer 40 cm or more thick begins within the depths of the subsurface tier.

Dysic—pH is less than 4.5 (in 0.01*M* $CaCl_2$) throughout all parts of the subsurface tier or the surface and subsurface tiers if a mineral layer 40 cm or more thick begins within the depths of the subsurface tier.

TEMPERATURE CLASSES

Same temperature classes as for mineral soils.

SOIL DEPTH CLASSES

Shallow—Lithic contact between 18 and 50 cm.
Micro—Lithic contact shallower than 18 cm.

SUMMARY

The classification of organic soils (Histosols) differs from the former systems in that it is based on quantitative criteria and criteria that can be determined in the field by visual observations and by simple field tests. Criteria used at the suborder level is principally degree of decomposition; great group level is principally soil temperatures; and at the subgroup level intergrades to other great groups of organic soils, the presence of a mineral or limnic layer, and presence of rock or permafrost or open water are recognized if any of these materials occur within the control section. The subgroup can further be separated into families using mineralogy for the terric subgroups, kind of limnic sediment for limnic subgroups, reaction, soil temperature, and soil depth.

LITERATURE CITED

Farnham, R. S., and H. R. Finney. 1965. Classification and properties of organic soils. Advanc. Agron. 17:115-162.

Soil Survey Staff. 1974. Soil taxonomy: A basic system of soil classification for making and interpreting soil surveys. USDA Agriculture Handbook No. 436. U. S. Govt. Printing Office, Washington, D. C. (In press).

Field Laboratory Tests for Characterization of Histosols[1]

2

W.C. LYNN, W.E. MC KINZIE, and R.B. GROSSMAN[2]

ABSTRACT

Simple procedures to characterize organic soil material in a field office or similar setting with a minimum of equipment are outlined. A half-syringe measuring device is used to determine fiber content by volume. A total of 183 samples from coastal marshes and swamps and from inland bogs were tested. Assessment of state of decomposition by physical (rubbed fiber content) and chemical (pyrophosphate solubility) measurement is consistent and generally in agreement with field estimates. Data for state of decomposition, mineral content, and bulk density are compared in detail.

INTRODUCTION

This paper describes procedures to characterize organic soil material for the purpose of classifying soils in the USDA Soil Taxonomy System. Relationships among the state of decomposition, mineral content, and bulk density are examined in detail. The state of decomposition is assessed physically by content of resistant fiber and chemically by solubility in sodium pyrophosphate. The procedures may be performed in the field office or similar setting with simple equipment.

METHODS

Place a representative sample of organic material into a 60-ml (2 ounce) plastic container for interim storage. If the sample is dry or nearly dry, add water to the container and allow to stand for 1 day or longer. Transfer the sample to a paper towel, roll the towel around the sample, and blot externally with additional paper towels to remove excess moisture. Squeeze gently to insure firm contact between the towel and sample. Unroll the towel and cut the cigar-shaped residue into approximately 1-cm sections. The prepared sample generally has a bulk density between 0.18 and 0.30 g/cc and a water content between 300 and 500%.

[1] Contribution from Soil Survey Investigations and Soil Correlation Units, Midwest Technical Service Center, SCS, USDA.

[2] Soil Scientists, Soil Conservation Service, USDA, Lincoln, Nebraska.

For determining fiber content, pyrophosphate solubility, or pH, pack pieces of the prepared sample into a 5-cc half-syringe adjusted for a volume of 2.5 cc (a 5-cc plastic syringe is cut in two, longitudinally, to make the half-syringe). In packing the half-syringe, compress the sample just enough to saturate the material and force out any entrapped air. Do not express any water. It is to this moisture condition that the residue must be returned later, when the residue volume is determined.

For fiber determination, transfer the 2.5-cc sample to a 100-mesh sieve and wash under running tap water until the effluent appears clear. Remove excess moisture through the sieve from the underside by blotting with an absorptive tissue. Repack the residue into the half-syringe, and blot further with an absorptive tissue until the moisture content reaches the state described above. Read the residue volume on the half-syringe and record it as percent, unrubbed fiber. Transfer the residue to the 100-mesh sieve and rub between the thumb and forefinger under a stream of running tap water until the effluent is clear. Blot and repack the residue into the half-syringe as for the unrubbed fiber. Read the volume and record it as percent rubbed fiber.

For determination of solubility in a saturated pyrophosphate solution, mix a 2.5-cc half-syringe sample with 1/8 teaspoon (about 1 g) of sodium pyrophosphate crystals and 4 ml of water in a 30-ml plastic container, and allow the mixture to stand overnight. Mix again and insert a strip of chromatographic paper (1/2 by 3 cm) to absorb the colored solution. Allow the strip to moisten completely. Tear off the soiled end, blot the strip gently on another sheet of chromatographic paper, and compare the colored strip with a Munsell color chart.

To determine pH, mix a 2.5-cc half-syringe sample with 4 ml of 0.015*M* $CaCl_2$ and equilibrate for at least 1 hour. Determine the pH by glass electrode or by pH paper.

Water content, mineral content, and bulk density can be determined on a single sample. Trim a core of undisturbed organic fabric to fit snugly into a tared (T), 130-ml aluminum moisture can. Take care not to disrupt the sample internally. The can volume (V) is used for bulk density. Cover tightly and weigh (A) as soon as practical. Dry the sample overnight at 110C, cool and weigh (B). Heat the sample overnight at 400C, cool, and weigh (C). A small, 110V muffle furnace is available that can be used in the field office. The pertinent computations follow:

$$\text{Water content (\%)} = (A - B)/(B - T) \times 100 \qquad [1]$$

$$\text{Mineral content (\%)} = (C - T)/(B - T) \times 100 \qquad [2]$$

$$\text{Bulk density (g/cc)} = (B - T)/V \qquad [3]$$

Water content can be expressed on the basis of field weight $[(A - B)/(A - T) \times 100]$ or on the basis of field volume $[(A - B)/(V) \times 100]$. Moisture contents at other specified tensions can be determined in the laboratory.

MATERIALS

A total of 183 samples were analyzed in conjunction with field sampling and study trips to Louisiana, Minnesota, and eastern Canada, and on samples from Minnesota, Wisconsin, Michigan, Indiana, Illinois, and North Dakota submitted to the Soil Survey Investigations Unit in Lincoln, Nebraska. The Louisiana samples were taken from coastal marshes and swamps. The samples from the northern states include a variety of materials, largely from cultivated bogs. The Canadian samples were taken from wild bogs, cultivated bogs, and commercial peat bogs and, by field evaluation, included a wide range of decomposition states.

RESULTS AND DISCUSSION

One objective of the study is to compare independent estimates of state of decomposition obtained via physical and chemical approaches. To facilitate graphical analysis, a pyrophosphate index (PI) was derived from the Munsell color notation by subtracting the chroma from the value. Figure 1 indicates PI values for each color chip of a 10YR Munsell page. For taxonomic purposes, a PI of 5 or more indicates fibric material, and a PI of 3 or less in-

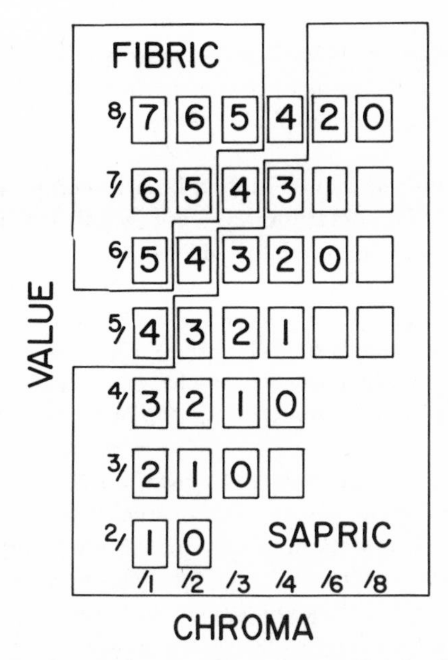

Figure 1. Copy of a 10YR page from a Munsell color book with pyrophosphate index (value minus chroma) indicated for each color chip. Areas corresponding to fibric and sapric materials are outlined.

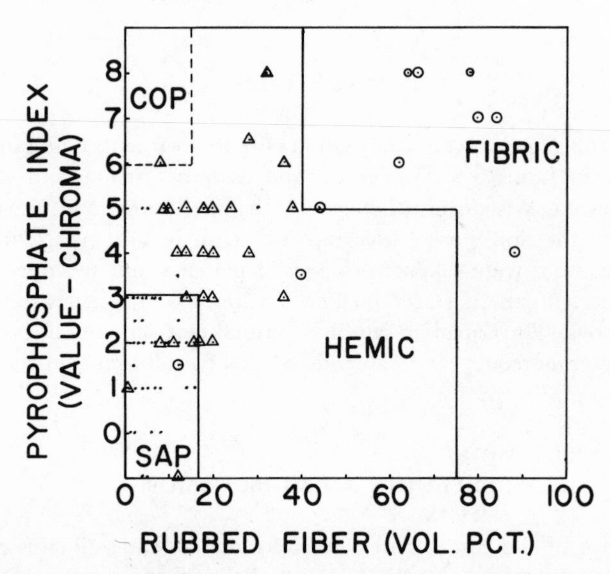

Figure 2. Results of field laboratory tests for pyrophosphate index and rubbed fiber percentage. Solid lines delineate areas of fibric, hemic, and sapric materials. The dotted line suggests possible boundaries for limnic (coprogenous) material. The type of symbol denotes the field evaluation of state of decomposition; ⊙ = fibric, △ = hemic, ● = sapric.

dicates sapric material. This differs from the present taxonomic criteria by including the 6/1 color chip with fibric materials and color chips with a chroma of 5 or more and a value of 8 with sapric materials. To date, none of the test results include pyrophosphate colors for any of the reassigned color chips.

The PI can be plotted against rubbed fiber to compare the chemical and physical approaches for determining state of decomposition. Present taxonomic criteria require a material to meet both the rubbed fiber and pyrophosphate solubility requirements to qualify for sapric or for fibric. Other materials are hemic. Areas corresponding to sapric, hemic, and fibric materials are separated by solid lines in Figure 2. The area bounded by a dotted line in the upper left corner suggests limits for coprogenous (limnic) material. Data in the paper by H. R. Finney et al., (chapter 3, this book) support the boundaries outlined.

The data points in Figure 2 tend upward and to the right, indicating consistency between the two measures of degree of decomposition. Many of the data points represent determinations on several samples. The symbol used to represent a data point indicates the field assessment of the state of decomposition. There is fairly good agreement between test results and field assessment. It should be noted that correlation between test data and the field descriptive assessment has improved as personnel have gained experience.

The rubbed fiber content was utilized to characterize the state of decomposition for comparison with other properties. Rubbed fiber assessment

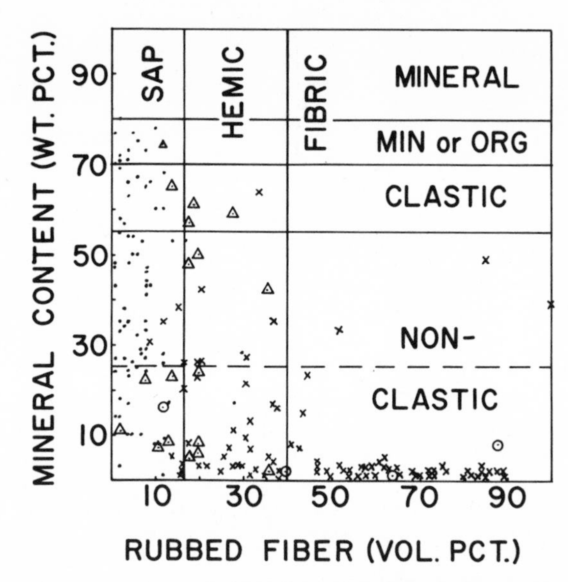

Figure 3. Mineral content and state of decomposition. Vertical boundaries separate materials by state of decomposition according to rubbed fiber criteria. Horizontal boundaries separate taxonomic classes according to mineral content. Commercial peat moss must have less than 25 percent material as indicated by the dashed line (ASTM, 1969); ⊙ = fibric, ◬ = hemic, ● = sapric; x = data from Nowland, 1971.

corresponds closer to daily field mapping operations than does the pyrophosphate solubility test.

The state of decomposition, i.e., rubbed fiber content, is commonly associated with mineral content of the soil. As fiber content decreases, the mapper suspects the possibility of a higher mineral content. The relationship between rubbed fiber and mineral content for the test data is shown in Figure 3. Areas in the graph are divided according to taxonomic classes for state of decomposition and mineral content. Few samples contain both high rubbed fiber and high mineral content. The mineral contents of only three fibric materials exceed the 25% limit set for commercial peat moss (ASTM, 1969). All of the fibric materials are sphagnum mosses, and most have mineral contents below 5%. Some hemic materials have intermediate mineral contents but none have more than 65% mineral. The mineral content of sapric materials ranges widely from 1 to 78%.

The mineral material is heavy compared to the organic material and has a significant effect on the bulk density of a given sample. The relationship between soil bulk density and mineral content is shown in Figure 4*a*. Taxonomic classes are indicated according to mineral content. Bulk density trends upward only slightly below a mineral content of 65%, and rises very rapidly above a mineral content of 85%. The transition zone between mineral contents of 65 and 85% corresponds roughly to the taxonomic transition between organic and mineral soils.

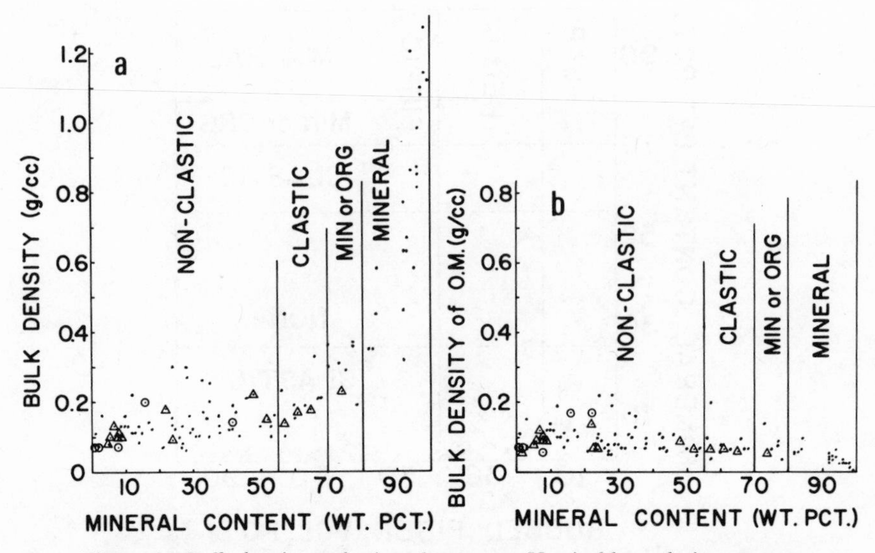

Figure 4. Bulk density and mineral content. Vertical boundaries separate taxonomic classes according to mineral content. (*a*) Soil bulk density as a function of mineral content. (*b*) Organic bulk density (weight of mineral computed out) as a function of mineral content. ⊙ = fibric, △ = hemic, ● = sapric.

The weight contribution of the mineral component can be calculated out of the bulk density by Eq. [4],

$$Db_{OM} = Db - Db(WT\% \; MIN/100) \qquad [4]$$

where Db_{OM} is the bulk density of the organic framework, Db is the bulk density of the soil, and WT% MIN is the weight percent of mineral material. This permits examination of relationships between bulk density of the organic component and various other soil properties. Computations are based on the important premise that the organic soil fabric is an open, skeletal framework of organic material in which mineral matter occupies interstitial space but does not alter the bulk volume.

Bulk density of the organic framework is plotted against mineral content in Figure 4*b*. The trend is for bulk density to remain constant or to decrease slightly as mineral content increases from 0 to 85%. These data support the concept of a skeletal organic framework, and suggest the concept is valid below a mineral content of 85%. Bulk densities above 0.15 g/cc for the organic framework are generally for samples from cultivated surface horizons. Bulk density of the organic component averages around 0.07 to 0.08 g/cc (see Figure 6*b* also).

The disparity in bulk densities between the mineral and organic fraction has an important bearing upon estimation of mineral content in the field. In an organic soil, the mineral component may constitute a significant propor-

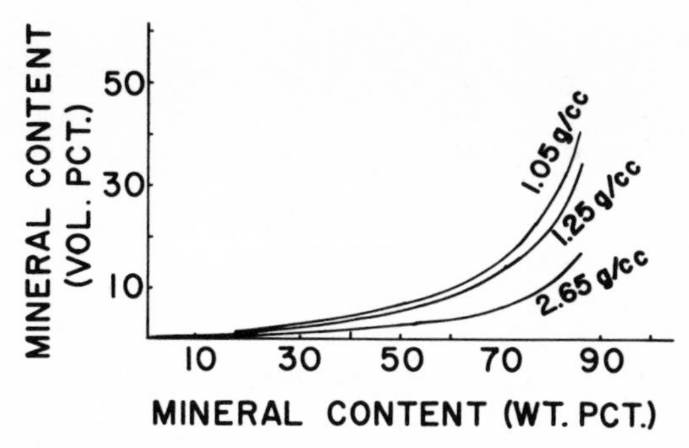

Figure 5. Comparison of mineral content expressed as a volume percent and as a weight percent. Taxonomic classes are outlined according to weight percent criteria.

tion of the weight but only a few percent of the volume. The relationship is outlined in Figure 5 for three densities of mineral material. At the boundary between clastic and nonclastic taxonomic families, i.e., 55% mineral by weight, the mineral component would occupy between 3% (for quartz sand) and 8% (for dispersed clay) of the soil volume. Because of this weight-volume relationship, there is a serious question whether the distinction between nonclastic and clastic families can be made reliably in field mapping operations. The problem is compounded in sapric materials because it is difficult to distinguish clayey mineral material from humified organic material. Subsequent to the present study, clastic families were deleted from *Soil Taxonomy* (Soil Survey Staff, 1974).

Data thus far have shown that although soil bulk density of a given sample is affected by mineral content, the bulk density of the organic framework is affected little by the mineral content. Next, the effect of state of decomposition upon bulk density is examined in a similar fashion. The relationship between soil bulk density and rubbed fiber is shown in Figure 6*a*. Sapric, hemic, and fibric materials are outlined according to taxonomic criteria for rubbed fiber. Soil bulk densities range from 0.05 to 0.15 g/cc for all the fibric and most of the hemic samples. Bulk densities range more widely for the sapric materials, related largely to the effect of mineral content. Bulk densities above 0.25 g/cc are limited to materials with less than 7% rubbed fiber, and most of these are from cultivated surface horizons.

Figure 6*b* shows the relationship between the state of decomposition and the bulk density of the organic framework. Only in the sapric range do bulk densities of the organic framework exceed 0.15 g/cc. The plot indicates a slight decrease in bulk density of the organic framework with increased fiber. The more important observation, however, is that the bulk density of the organic framework is little affected by the state of decomposition of the organic matter.

The bulk density and mineral content can be utilized directly in an assessment of subsidence potential for organic soils. Organic materials undergo

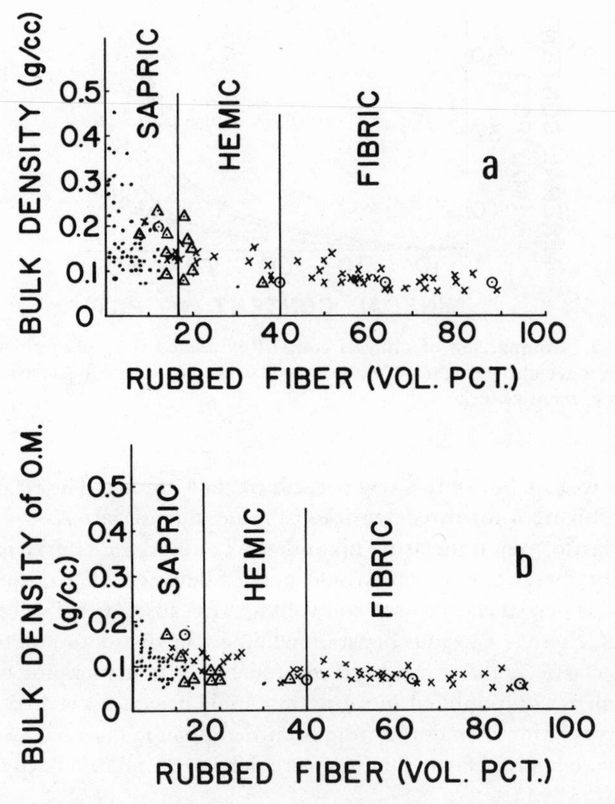

Figure 6. Bulk density and state of decomposition. Vertical boundaries separate fibric, hemic, and sapric materials according to rubbed fiber criteria: (*a*) Soil bulk density as a function of rubbed fiber content. (*b*) Organic bulk density (weight of mineral computed out) as a function of rubber fiber content. ⊙ = fibric, △ = hemic, ● = sapric; x = data from Nowland, 1971.

initial subsidence or consolidation soon after drainage, followed by continued subsidence from oxidation and dissipation of the organic materials. Losses in elevation from continued subsidence generally range from 1 to 5 cm/year. If the entire organic component is lost, the residue, termed here the minimum residue, reflects the mineral component of the original material. The minimum residue depends on the initial bulk density, the mineral content, and the bulk density of the residue.

$$\text{Minimum residue (cm/cm)} = \frac{\text{Db}_{\text{initial}} \times \dfrac{\text{\% Mineral by wt}}{100}}{\text{Db}_{\text{residue}}} \quad [5]$$

Initial bulk density is the bulk density at the time of sampling, which may be before or after the subsidence process has started. By assuming rea-

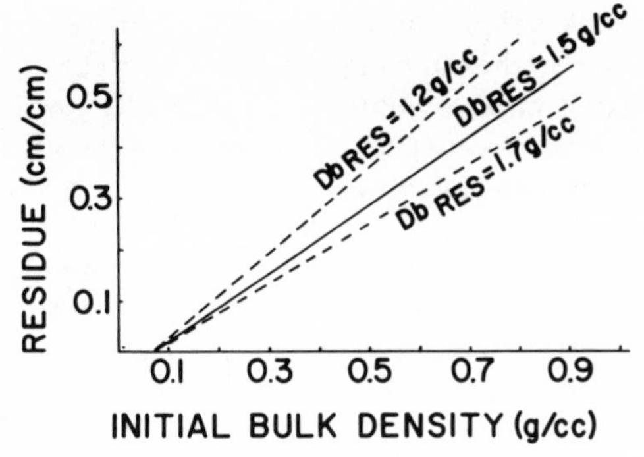

Figure 7. Relationship between the minimum residue after subsidence and the initial soil bulk density.

sonable bulk density for the organic framework (0.07 g/cc, Figures 4*b* and 6*b*) and the residue (1.5 g/cc), the relationship can be simplified.

$$\text{Minimum residue (cm/cm)} = \frac{Db_{\text{initial}} - Db_{\text{OM}}}{Db_{\text{residue}}} \qquad [6]$$

$$= \frac{Db_{\text{initial}} - 0.07}{1.5} \qquad [7]$$

Equation [7] was utilized to plot minimum residue as a function of initial bulk density in Figure 7. The *x*-intercept is the bulk density of the organic framework and the slope depends on the residue bulk density. To estimate the thickness of a residue that contains organic matter, reduce the residue bulk density appropriately.

SUMMARY

Procedures are outlined to characterize organic materials in the field office or laboratory with simple equipment and techniques. Determinations are detailed for unrubbed and rubbed fiber, solubility in saturated sodium pyrophosphate, pH, water content, mineral content, and bulk density. Fiber contents are determined as volume percents by means of a half-syringe measuring device. A pyrophosphate index was developed to facilitate comparison of pyrophosphate solubility data. A total of 183 samples from coastal marshes and swamps, and from inland bogs were tested.

Evaluation of state of decomposition by rubbed fiber and pyrophosphate solubility is in reasonable agreement, and the tests are fairly reliable. Bulk density of the organic component is little affected by the state of decomposition or mineral content, and is generally restricted to the range between 0.05 and 0.15 g/cc for samples with less than 85% mineral material. These data support the premise that the organic fabric is an open, skeletal framework with bulk volume little affected by mineral content. Soil bulk densities are generally in the same range (0.05-0.15 g/cc) for fibric and hemic materials because the mineral contents are generally low. Sapric materials range to higher mineral contents, and therefore bulk densities range to higher values. Cultivated surface horizons tend to have higher bulk densities than uncultivated surface horizons. A substantial mineral content by weight equates with only a few percent mineral matter by volume. The maximum subsidence or the minimum residue for drained organic soils can be determined from the initial bulk density and reasonable assumptions for bulk densities of the organic framework and the residue.

LITERATURE CITED

American Society for Testing and Materials. 1969. Standard D2607-69. Standard classification of peats, mosses, humus, and related products. American Society for Testing and Materials. Philadelphia, Pa.

Nowland, J. L. 1971. Organic soils tour (eastern Canada)–Nova Scotia section. Nova Scotia Agricultural College, Truro.

Limnic Materials in Peatlands of Minnesota[1]

3

H.R. FINNEY, E.R. GROSS, and R.S. FARNHAM[2]

ABSTRACT

Limnic materials (post-glacial lake sediments) were observed to underlie a significant portion of the peatlands of Minnesota. The purpose of this investigation was to measure the properties of these materials that are important in characterizing them as soil materials. About 40 pedons were investigated in peatlands in the major soil-geomorphic areas of Minnesota. The thickness of peat overlying the limnic materials ranged from less than 1 m to more than 5 m. Limnic materials varied widely in morphologic features, especially in regard to color and content of plant detritus. Bulk density ranged from about 0.1 to about 0.9 g/cc, and the content of mineral material ranged from about 10 to about 90%. The coefficient of linear extensibility (COLE) values ranged from 5 to about 120%. Generally the content of mineral material and bulk density increased with depth and the COLE values decreased with depth. The coprogenous earth and marl classes of limnic material each included materials with a wide range of properties. Each body of peatland usually had limnic materials unique in certain respects, especially the stratigraphic sequence of properties. Thus, we conclude that the possibility of developing models for predicting the nature of these materials on a regional basis is remote.

INTRODUCTION

When the eutrophification of lakes has reached the stage that the limnic materials[3] (post-glacial lake sediments) are covered with a mantel of peat[4], the further study of both materials comes within the realm of the pedologist. Limnologists [see contributions from the Limnological Research Center, University of Minnesota, Swain (1967), for example] have studied Minnesota limnic materials in some detail but have not studied to any extent certain

---

[1]Paper no. 1502 of the Scientific Journal Series, Minnesota Agricultural Experiment Station, University of Minn., St. Paul, Minnesota.

[2]Soil Correlator, Soil Conservation Service, USDA, and Research Assistant and Professor of Soil Science, University of Minnesota, respectively, St. Paul, Minnesota.

[3]Limnic materials, as defined in *Soil Taxonomy* (Soil Survey Staff, 1974) are materials that are deposited in lakes and, that consist primarily of chemical and biological precipitates or slightly to mostly decomposed aquatic organisms or both.

[4]Peat as used here is synonomous with sapric, hemic, and fibric kinds of organic soil material as defined in *Soil Taxonomy* (Soil Survey Staff, 1974).

properties that are important in characterizing them as soil materials. Chief among such properties are bulk density, content of water at various suctions, and shrinkage upon drying. Pedologists in Minnesota have done little work on the study of these materials until the past few years.

The fact that limnic materials underlie significant areas of peatlands of Minnesota has been recognized for some time (Soper, 1919). However, earlier soil classification systems of the USA did not specifically recognize such materials: areas of peatlands with limnic materials at shallow depths were not distinguished from areas where the overlying peat was thicker. One special exception, however, was the recognition of separate taxa at lower levels of abstraction for limnic materials that were exposed in drained shallow lakes.

The present system of soil classification (*Soil Taxonomy,* Soil Survey Staff, 1974) provides taxa for organic soils with limnic materials if they are at shallow depths. Further, it gave considerable impetus to the investigation of peatlands in Minnesota and elsewhere in the USA. Some results of our recent studies of limnic materials in peatlands of Minnesota are reported herein.

DISTRIBUTION AND EXTENT

The peatlands of Minnesota comprise about 2.6 million ha (6.5 million acres). We estimate that limnic materials underly about 30% of the peatlands. Such peatlands primarily are in glacial moraines that have irregular relief with many depressions and lakes. Figure 1 shows the distribution of the areas of Minnesota containing peatland and areas where peatlands with limnic materials are dominant. Recent soil surveys in some areas of southern Minnesota indicate that peatlands with limnic materials at depths of 1.5 m or less comprise about 70% of the total. Further, where lakes were drained in the southern part of the state, limnic materials were at the soil surface. Some of these materials are being cultivated.

INVESTIGATION PROCEDURES

During the past few years we have described and sampled in detail about 40 pedons containing at least some limnic materials. Also, many other pedons have been studied in lesser detail. Most samples were taken using an auger that was designed by the Macaulay Institute for Soil Research, Aberdeen, Scotland. It has a barrel about 50 cm in length and about 5 cm in diameter. It is the best auger we have found for obtaining undisturbed samples. With it, samples were obtained to be described and analyzed. For the laboratory analyses, we obtained bulk samples, samples of known volume for the determination of bulk density, and undisturbed pieces of material that are

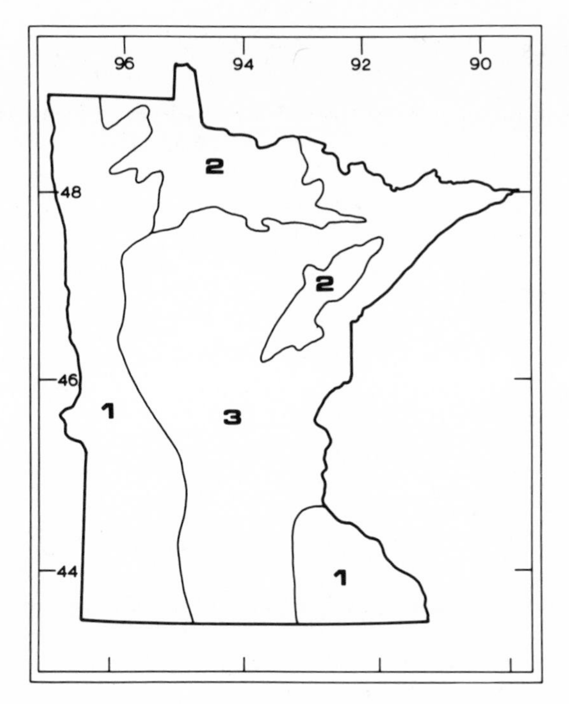

Figure 1. The distribution of peatlands in Minnesota. Areas 1 have essentially no peatlands. Areas 2 have vast expanses of peatlands, but most peatlands lack underlying limnic materials. Area 3 has moderate amounts of peatland, and most peatlands have underlying limnic materials.

coated with Saran for subsequent determination of bulk density and water retention.

In the field we quantitatively described the following features of limnic materials:

1) Depth and thickness of layers;
2) Color of the broken face of pieces and a rubbed mass;
3) Contents and nature of plant detritus, mineral material, and shells of mollusks;
4) Structure;
5) Consistence;
6) Reaction and effervescene with dilute HCl; and
7) Boundary.

We measured the following properties in the laboratory.

MINERAL MATERIAL

Estimates of the content of mineral material are made by firing the sample at 375C for 16 hours in a muffle furnace (Ball, 1964). Contents are reported as percent of oven-dry weight.

PLANT DETRITUS

Plant detritus in limnic materials is defined as pieces of plant tissue that are large enough to be retained on a sieve with openings of 0.15 mm. Samples are mixed in a solution of 0.025M sodium pyrophosphate for about 10 min, left undisturbed for 12 hours, and then washed on the sieve. Contents are reported as percent of oven-dry weight of sample.

WATER CONTENTS

Water contents at 1/3-bar suction are measured on saran-coated fragments and at 15 bar from a bulk sample, both with the pressure-plate extraction method (Soil Survey Staff, 1967).

FREE CARBONATES

Free carbonates are measured by treating the samples with HCl and measuring the amount of CO_2 evolved (Soil Survey Staff, 1967). Results are reported as calcium carbonate equivalent percent oven-dry weight.

BULK DENSITY

Bulk density is measured on the basis of volume at saturation, at a suction of 1/3 bar, and at oven-dry. The former is on bulk samples of known volume and the latter two measurements are on saran-coated pieces.

SOLUBILITY IN SOLUTION OF SODIUM PYROPHOSPHATE

The solubility of the organic fraction in a solution of sodium pyrophosphate ($Na_4P_2O_7$) indicates the degree of oxidation or humification (Mackenzie and Dawson, 1962). A sample of material is placed in a solution of $Na_4P_2O_7$, which contains an excess of solute, and is mixed and allowed to stand for about 16 hours. A piece of chromatographic paper is inserted into the mixture and the color of the material absorbed is measured. This procedure "helps" to distinguish dark-colored limnic materials from sapric or hemic classes of organic soil material (Soil Survey Staff, 1974).

COEFFICIENT OF LINEAR EXTENSIBILITY

The coefficient of linear extensibility (hereafter referred to as COLE) is a measure of the shrinkage of materials. It is computed as follows (Soil Survey Staff, 1967):

$$\mathrm{COLE} = \left(\frac{Db_d}{Db_m}\right)^{1/3} - 1$$

where Db_d is the bulk density on the basis of volume when oven dry and weight when oven dry and Db_m is the bulk density on the basis of the volume at 1/3-bar suction and oven-dry weight. Saran-coated pieces are used in this determination.

CATION EXCHANGE CAPACITY

The cation exchange capacity is measured using NH_4OAc as a displacing compound. The procedure is patterned after method *5Alb* of the Soil Survey Staff (1967) except that KCl is used instead of NaCl and isopropyl alcohol is used instead of ethanol.

RESULTS AND DISCUSSION

Properties

We found a wide range in both depth to and thickness of limnic materials. We observed them, at one extreme, under 5.6 m of peat and at the other extreme, at the surface. The latter, however, is rather unique because such areas formerly were shallow lakes that have been drained within the past century. Also, a lake that is generating limnic materials still has open water and such limnic materials are not considered to be soil in the conventional sense. The thickness of limnic materials ranged from as little as 0.3 m to as much as 5.7 m.

Three kinds of limnic materials are recognized in the *Soil Taxonomy*. To facilitate the following discussion, their names and major distinguishing features (Soil Survey Staff, 1974) follow:

Coprogenous Earth—is a limnic layer that

1) Has a moist color value of less than 5 (Munsell color designation);
2) Forms a slightly viscous suspension in water and is slightly plastic when wet or shrinks on drying to form clods that are difficult to rewet and that commonly tend to crack along horizontal planes;
3) Yields saturated $Na_4P_2O_7$ extracts that are higher than 7 in value and lower than 3 in chroma (both Munsell color designations).

Coprogenous earth, for example, includes materials called "dy", "gyttja", and "sapropel" by Kubiena (1953) in Germany; "copropel", "sapropel", and some "peat" as well as many intergrades between those entities by Swain (1956, 1970) in Minnesota; and "sedimentary peat" by Rigg (1958) in Washington.

Marl—is a limnic layer that

1) Consists mostly of calcium carbonate;
2) Has moist color value of 5 or more (Munsell color designation).

Diatomaceous Earth—is a limnic layer that

1) Consists mostly of skeletons of diatoms;
2) Has a matrix color value of 3 through 5 (Munsell color designation) if not previously dried, and changes value irreversibly on drying;
3) Yields saturated $Na_4P_2O_7$ extracts that are higher than 7 in value and lower than 3 in chroma (both Munsell color designations).

We observed all of these kinds of materials in peatlands of Minnesota. However, coprogenous earth is the dominant limnic material. Marl is common where the glacial drift contains alkaline earth. Essentially it is lacking in the areas covered with drift of the Superior and Rainy lobes. Diatomaceous earth was observed in only one small area in southwestern Minnesota.

In peatlands containing marl, the common stratigraphic sequence from the surface is; peat, coprogenous earth, marl, a thin layer of coprogenous earth, and a mineral substratum of glacial drift. However, in some places peat directly overlies marl. Also, some peatlands show evidence of catastrophy in their development. For example, in some areas marl is over layers of peat and in others it is between layers of peat.

Coprogenous earth typically has color with hue of 2.5Y or 5Y, value of 3 or 4, and chroma of 1 or 2, but the full range in color comprises hue of 10YR through 5Y, value 2 through 4, and chroma of 1 or 2. Its content of plant detritus ranges from a trace to as much as 50% with a common range being 10 to 20%. Typically it is massive, but in some places it has weak, very fine or fine, granular structure. It typically is nonplastic and nonsticky but ranges to slightly sticky and slightly plastic. Its pH in 0.01*M* $CaCl_2$ ranges from 4.0 to 7.9 and it commonly lacks free carbonates, but contents as high as 60% have been measured.

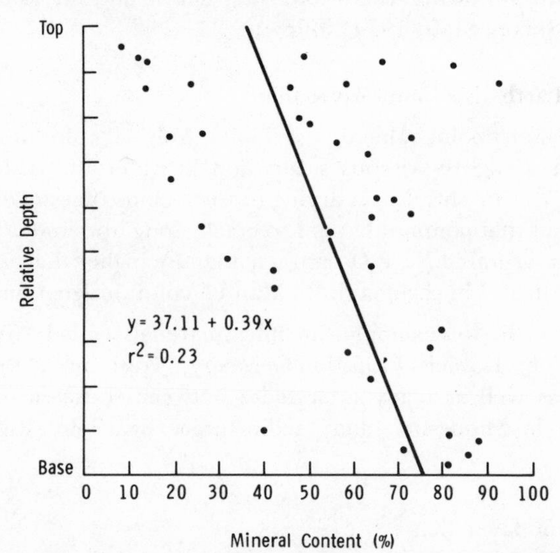

Figure 2. Content of mineral material in relationship to relative depth (in the limnic material) of some samples of coprogenous earth.

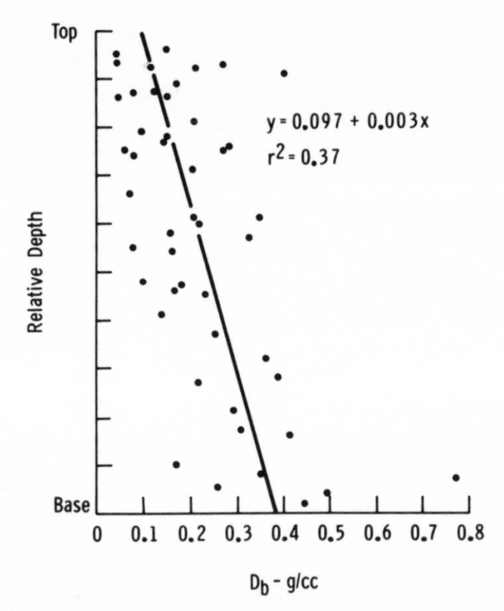

Figure 3. Bulk density in relationship to relative depth (in the limnic material) of some samples of coprogenous earth.

Relationships between relative depth and mineral content, relative depth and bulk density, mineral content and bulk density, and COLE and bulk density for coprogenous earth from several places in Minnesota are shown in Figures 2 to 5, respectively. The values for these features range widely; and relationships, especially between relative depth and mineral content and relative depth and bulk density, are weak. A stronger relation-

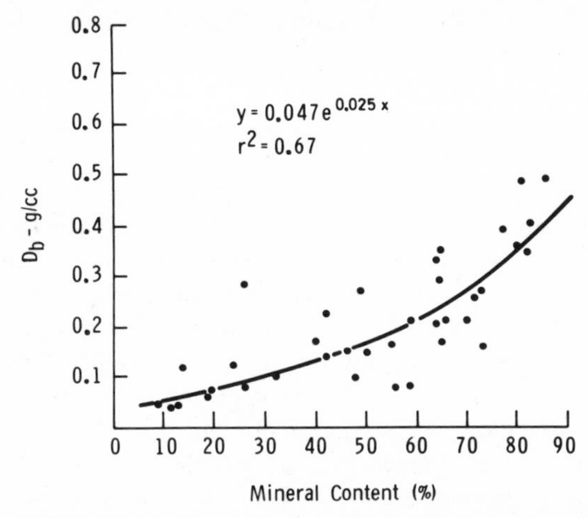

Figure 4. Relationship between bulk density and mineral content of some samples of coprogenous earth.

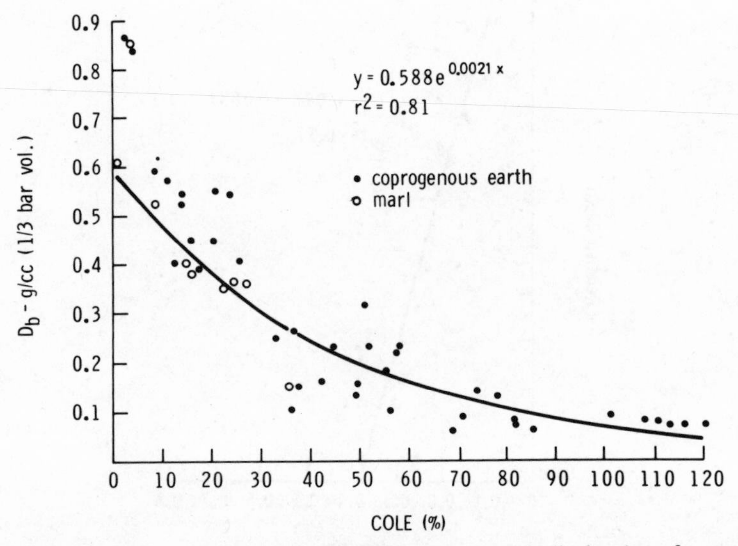

Figure 5. Relationship between COLE values and bulk density of some samples of limnic materials.

ship exists between mineral content and bulk density and between COLE and bulk density.

The location of marl deposits and their content of lime have been investigated thoroughly in Minnesota (see Thiel, 1933 and Schwartz et al., 1959). Contents of $CaCO_3$ (equivalent) ranged from 50 to 90% generally, but we have measured values as low as 18 to as high as 99%. Marl typically has color with hue of 10YR or 2.5Y, value of 5 or 6, and chroma of 1 or 2 with the full range comprising hue of 10YR through 5Y, value of 5 through 8, and chroma of 1 or 2. Its content of plant detritus typically was less than 5% but ranged to 15%. Shells of mollusks comprised as much as 50% of the mass in some places, but they were lacking in some marls. Values for bulk density and COLE are shown in Figure 5.

As in other properties, the solubility of the organic fraction in a solution of sodium pyrophosphate ranges widely both for coprogenous earth and for marl. For some materials that we would prefer to call coprogenous earth because of their genesis, extracts with color value and chroma as low as 3 have been measured. For others, a value as high as 8 and a chroma as low as 1 have been measured. A common range in color of extract was a value of 6 through 8 and a chroma of 1 or 2. As defined, materials that have an extract with value of less than 7 and chroma of more than 2 are excluded from coprogenous earth. Depending on content of organic matter, such materials are either sapric organic soil material or mineral soil material. Further, the color of extracts of marl ranged primarily from 4 through 8 in value and from 1 through 3 in chroma. Values of 7 or 8 were most common.

The content of water at various suctions ranged widely in limnic materials. Values at a suction of 1/3 bar ranged from 60 to 1,200% (oven-dry weight basis), and values at 15 bar ranged from about 30 to 170%. Further, for most samples, water retained at 1/10 bar was only slightly more than that retained at 1/3 bar. Further analysis of water retention is needed before relationships between this and other properties can be tested.

Cation exchange capacity was determined on only 14 samples of limnic materials. Values ranged from 17 to 77 meq/100 g with values of 40 to 55 meq/100 g being most common. All of these samples lacked free carbonates.

CLASSIFICATION

Most limnic materials are mantled with peat, and are placed in the order of Histosols at the subgroup and lower categories. Their presence only within the control section (within a depth of 130 cm, or 160 cm in some situations) is the criterion at the subgroup category. For example, the subgroup of Limnic Borofibrists includes Borofibrists that have limnic materials within the control section. Three kinds of limnic material are recognized at the family category, namely, marl, coprogenous earth, and diatomaceous earth. Thus, a coprogenous family class of Limnic Borohemists has a limnic material dominated by coprogenous earth within the control section.

Soil series in limnic subgroups of Histosols identified in soil surveys in Minnesota are shown in Table 1. All of these series have been defined or introduced into Minnesota in the past 5 years. The descriptions reflect an improvement in our understanding and mapping of peatlands. However, the coprogenous earth that presently is included in these series has a wide range in properties. Perhaps the definitions of these series need to be narrowed and others recognized.

We are using two soil series that consist entirely of coprogenous earth, namely the Blue Earth and Urness series. Their classification is not yet settled. As generally used, these series include materials with a content of mineral material that ranges from 65 to 90%. Thus, they comprise both materials that qualify as organic soil material as well as mineral soil material. These materials show essentially no impress of soil genesis. Thus, some qualify for the order of Entisols, perhaps in the great group of Fluvaquents.

Table 1. Soil series of Histosols in limnic subgroups that currently are being used in soil surveys in Minnesota and their placement at the family category.

Series	Family
Metogga*	Limnic Medifibrists, coprogenous, euic, mesic
Millerville	Limnic Borohemists, coprogenous, euic
Carlos	Limnic Borohemists, marly, euic
Caron	Limnic Medihemists, coprogenous, euic, mesic
Rondeau	Limnic Borosaprists, marly, euic
Edwards	Limnic Medisaprists, marly, euic, mesic
Muskego	Limnic Medisaprists, coprogenous, euic, mesic

* Tentative series.

The remainder qualify for Histosols, but no lower category has been provided for their classification in *Soil Taxonomy* (Soil Survey Staff, 1974).

CONCLUSIONS

Limnic materials underlie significant areas of peatland in Minnesota. These materials vary widely in their properties. Only three kinds are recognized in the *Soil Taxonomy* (Soil Survey Staff, 1974) of the USA in contrast to many kinds recognized by limnologists. However, the three kinds are probably adequate for the family category. Additional distinctions can be made at the series category. Considering the wide variation in properties that are important in making interpretations, a great need exists for carefully measuring the nature of these materials in a soil survey.

A few soils consist entirely of limnic materials, and they are rather unusual. They pose some problems in classification because they are excluded or seem out of place in the taxon. A new taxon needs to be developed for them or other taxa need to be redefined. Also, the definition of coprogenous earth, especially in regard to degree of humification as indicated by solubility in sodium pyrophosphate, needs further consideration. Perhaps the requirements should be lowered for color value and increased for chroma. Considerably more study of limnic materials is needed before their classification can be considered firm.

Each lake is considered to have an environment unique in certain respects. Thus, the limnic materials that are generated in a given lake are somewhat different, especially in stratigraphic sequence of properties, from those in other lakes (or former lakes). Thus, the possibility of developing precise models for predicting the nature of these materials in peatlands on a regional basis is remote.

LITERATURE CITED

Ball, D. F. 1964. Loss-on ignition as an estimate of organic matter and organic carbon in non-calcareous soils. J. Soil Sci. 15:84–92.

Kubiena, W. L. 1953. The soils of Europe. Thomas Murby & Company, London. 317 p.

Mackenzie, A., and J. Dawson. 1962. A study of organic soil horizons using electrophoretic techniques. J. Soil Sci. 13:160–166.

Rigg, G. B. 1958. Peat resources of Washington. Dep. of Cons. Div. of Mines and Geol. Bull. 44. 272 p.

Schwartz, G. M., et al. 1959. Investigation of the commercial possibilities of marl in Minnesota. Office of Commissioner of Iron Range Res. and Rehabil., St. Paul, Minn. 123 p.

Soil Survey Staff. 1967. Soil survey laboratory methods and procedures for collecting soil samples. Soil Surv. Invest. Report No. 1, Soil Conservation Service, USDA, U. S. Govt. Printing Office, Washington, D. C. 50 p.

Soil Survey Staff. 1974. Soil Taxonomy: A basic system of soil classification for making and interpreting soil surveys. Soil Conservation Service, USDA Agriculture Handbook No. 436. U. S. Govt. Printing Office, Washington, D. C. In press.

Soper, E. K. 1919. The peat deposits of Minnesota. Minn. Geol. Surv. Bull. 16. 215 p.

Swain, F. M. 1956. Stratigraphy of lake deposits in central and northern Minnesota. Bull. Amer. Ass. Petrol. Geol. 40:600–653.

Swain, F. M. 1967. Stratigraphy and biochemical paleontology of Rossburg peat (recent), north-central Minnesota. Univ. Kansas Press, R. C. Moore Commemorative Volume. p. 445–475.

Swain, F. M. 1970. Non-marine organic geochemistry. (Cambridge Earth Science Series) Cambridge Univ. Press, London. 445 p.

Thiel, G. A. 1933. The marls of Minnesota. p. 79–190. *In* C. R. Stauffer and G. A. Thiel (ed.) The limestones and marls of Minnesota. Minn. Geol. Surv. Bull. 23.

The Hydrologic Characteristics of Undrained Organic Soils in the Lake States[1]

4

DON H. BOELTER[2]

ABSTRACT

Although the capacity of peat materials to take up and hold water has been recognized for many years, only recently have studies evaluated the physical and hydrologic characteristics of peat materials in their natural undrained profile. Water retention and hydraulic conductivity were found to vary greatly for different peat materials depending on their degree of decomposition. Although peatlands have sometimes been considered regulators of streamflow, the high water retention values and low hydraulic conductivities of moderately to well decomposed peats show that they give up too little water too slowly to maintain streamflow. Two-thirds of the annual streamflow from perched lake-filled bogs occurs prior to 1 June and they do not make any substantial contribution to streamflow during dry periods. The flat topography and high absorption capacity of the peat materials result in temporary short-term storage which reduces storm flow runoff peaks, particularly after summer drying periods when bog water tables are low.

INTRODUCTION

Undrained organic soils make up a significant part of the boreal and northern forest regions of North America. There are about 6 million ha of peatland in the Lake States of Michigan, Minnesota, and Wisconsin (Davis and Lucas, 1959). Approximately 75% of this peatland is considered forest or wildland and it makes up nearly 20% of the forested areas of these states. Few of these organic soils have been artificially drained. Instead, they are still characterized by water tables at or near the soil surface. These are the same wet conditions that limited decomposition of plant remains and resulted in their development in the first place.

These organic soils could play a significant role in amount and distribution of the water resources in the Lake States because they cover large portions of the headwaters catchment of most streams and rivers in the area. Only recently have studies been made to determine the properties of organic

[1]Contribution from North Central Forest Experiment Station, USDA Forest Service, Grand Rapids, Minnesota (Laboratory maintained in cooperation with University of Minnesota).

[2]Principal Soil Scientist, Northern Conifers Laboratory, Grand Rapids, Minn.

soils in their natural state. The present paper gives some results of these studies and reviews the hydrologic characteristics of organic soils in the peatlands of the Lake States.

PHYSICAL PROPERTIES OF ORGANIC SOILS

Water retention, water yield, and hydraulic conductivity are important physical properties of peat materials that determine to a large extent the hydrologic characteristics of organic soils. These properties are related to the porosity and pore size distribution which in turn are related to the degree of decomposition which can be estimated by bulk density and fiber content. As a group, peat materials are porous and hold large amounts of water when saturated. However, the nature of this porosity and the resulting values of physical properties vary considerably among different peat materials.

Water Retention

Organic soils in nature are wet not only because they exist under saturated conditions, but also because of their great affinity to take up and hold water. The high water retention of peat materials was reported by early investigators (Feustel and Byers, 1930; MacFarlane, 1959) and lead to the use of peat materials to improve the physical properties of soils used in greenhouses and nurseries (Feustel, 1938).

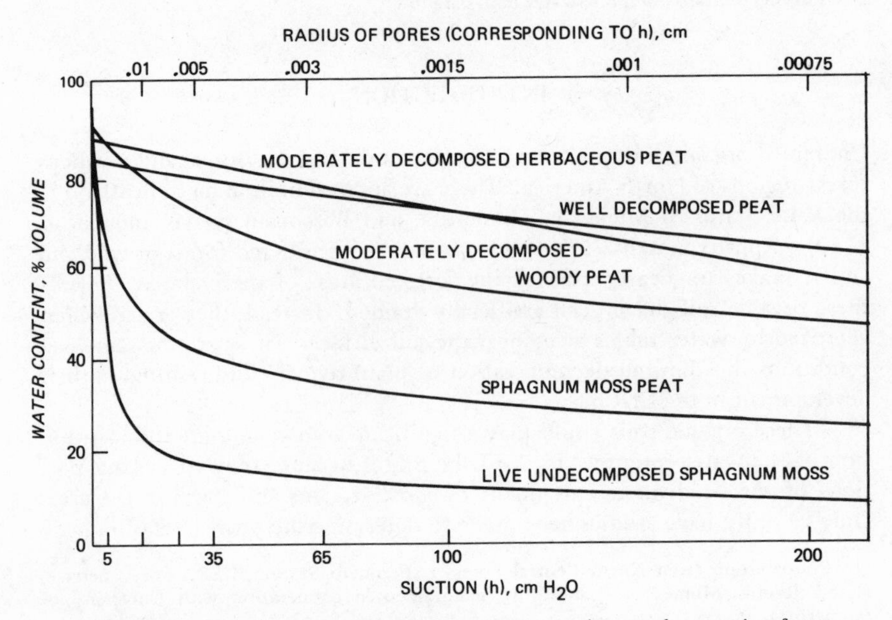

Figure 1. Relation between water content, suction, and pore size for several northern Minnesota peat materials.

The water content of saturated northern Minnesota peat varies from nearly 100% by volume for undecomposed sphagnum moss peat in surface or near surface horizons to about 80% by volume in more decomposed peat in deeper horizons (Boelter, 1969). Thus the total porosity decreases gradually with increasing decomposition, but is large for all peat compared to mineral soils.

Striking differences were found, however, in the quantity of water retained in peat at less than saturation, indicating the pore size distribution is more significant than total porosity. The undecomposed peats contain many large pores that are easily drained at low suctions. Water retention curves show that an undecomposed sphagnum moss peat loses a large portion of its saturated water content at suctions of only 5 cm of water (Figure 1). The finer pores of more decomposed peats do not drain at low suctions, and at suctions greater than 5 mbars, water contents are much higher than for the more undecomposed peat materials.

Water Yield Coefficient

Differences in water retention among organic soils are further illustrated with water yield coefficients, a measure of the quantity of water removed from a peat profile when the water table is lowered. These values include water removed from the saturated zone as well as water released from peat material in the capillary fringe but above the original zone of saturation.

Water yield coefficients of several northern Minnesota peat types were calculated from laboratory-measured water retention values (Table 1). The coefficients are equal to the differences in water content at saturation and

Table 1. Water yield coefficients, hydraulic conductivities, and bulk densities of several northern Minnesota peat materials (Boelter, 1970).

Peat type	Sampling depth	Hydraulic conductivity	Water yield coefficient	Bulk density
	cm	10^{-5} cm/sec	cc/cc	g/cc
Sphagnum moss peat				
Live, undecomposed mosses	0 - 10	--*	0.85	0.010
Undecomposed mosses	15 - 25	3810.00	0.60	0.040
Undecomposed mossed	45 - 55	104.00	0.53	0.052
Moderately decomposed with wood inclusions	35 - 45	13.90	0.23	0.153
Woody peat				
Moderately decomposed	35 -45	496.00	0.32	0.137
Moderately well decomposed	60 -70	55.80	0.19	0.172
Herbaceous peat				
Slightly decomposed	25 - 35	1280.00	0.57	0.069
Moderately decomposed	70 - 80	0.70	0.12	0.156
Decomposed peat				
Well decomposed	50 - 60	0.45	0.08	0.261

* Rates of water movement were too rapid to measure with the techniques used in this study.

0.1 bar suction and represent the change in water content with water level fluctuation in a profile consisting entirely of the specified peat type. The values range from 0.85 to 0.08 and are similar to values measured for bogs in Finland (Heikurainen, 1963). The values are related to pore size distribution and compare with values computed by Vorob'ev (1963) from pore volume distribution of peat samples from swamps in western Siberia. Any change in water table elevation in less decomposed peat, usually found in surface horizons, represents considerably more water than a corresponding change in deeper more dense peats.

Hydraulic Conductivity

Differences in pore size distribution of various peat materials also result in striking differences in the rate of saturated water movement. Hydraulic conductivities of various northern Minnesota peats (Boelter, 1965) were measured by piezometric methods in the field. A wide range of values were found (Table 1). Water movement was very rapid in surface or near surface horizons of undecomposed sphagnum moss peats. The more decomposed peat materials permitted very slow water movement with hydraulic conductivity values often lower than clays and glacial tills. Hanrahan (1954) and Colley (1950) both reported similar low hydraulic conductivity values for peat materials.

Bulk Density and Fiber Content

Because water retention, water yield coefficient, and hydraulic conductivity depend to a large degree on porosity and pore size distribution, they are, in turn, related to particle size distribution and structure. In peat materials, these properties are determined by the state of decomposition. Although the state of decomposition is difficult to quantify, both bulk density and fiber content have been used to estimate the degree of decomposition (Kaila, 1956; Farnham and Finney, 1965).

Numerous researchers have studied the decomposition of peats; however, only a few have reported quantitative comparisons with physical properties. Kuntze (1965) found that water retention characteristics of peat, such as field capacity and permanent wilting point, were dependent on decomposition as measured by bulk density, ash content, and humus content. Baden and Eggelsmann (1963) showed the relationship of hydraulic conductivity to substance volume and the von Post measure of decomposition.

Boelter (1969) showed there is a curvilinear relationship of water content at saturation (0 bar), 5 mbar, 0.1 bar, and 15 bar suctions, to unrubbed fiber content ($>$ 0.1 mm) and bulk density (Figure 2) with coefficients of multiple determination ranging from 0.66 to 0.88. Regression analysis of the logarithm of hydraulic conductivity on fiber content ($>$ 0.1 mm) and bulk density indicated a linear relationship ($r^2 = 0.54$).

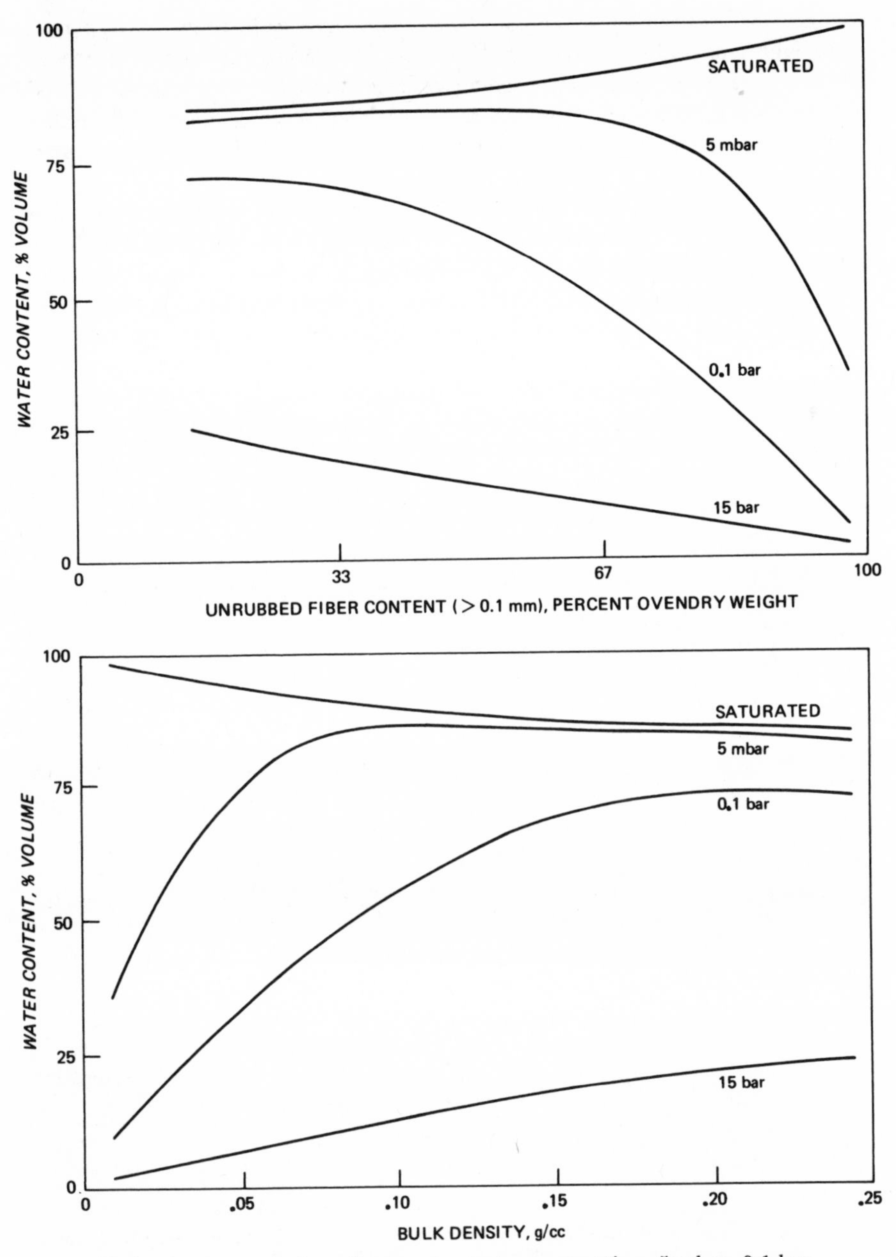

Figure 2. The relation of water contents at saturation, 5-mbar, 0.1-bar, and 15-bar suctions to unrubbed fiber content (> 0.1 mm) and bulk density (Boelter, 1969).

The nature of these relationships show that the most significant differences in physical properties occur with the least decomposed materials. The steepest area of most of the curves occurs at bulk densities less than 0.10 gm/

cc and fiber contents greater than 67%. The changes in physical properties with increased bulk density (or decreased fiber content) are not nearly as great with the more decomposed materials. Increased decomposition and accompanying changes in structure that occur with drainage (Okruszko, 1960), may ultimately improve the physical properties such as; water retention, water yield coefficient, and hydraulic conductivity.

Though relationships between physical properties and bulk density (or fiber content) are quite evident, woody peats seem to be different (Table 1). For a given bulk density, woody peats have a hydraulic conductivity and water yield coefficient similar to the less decomposed peats. It may therefore be necessary to consider woody peat separately.

The fiber contents used to develop the above relationships were unrubbed fiber contents using the Calgon dispersion technique. There is a growing concern about the reliability of this technique in estimating field unrubbed fiber contents.[3] Farnham et al.[4] indicated that these values lie between unrubbed and rubbed fiber contents and are probably closer to rubbed fiber contents.

The Calgon dispersion technique was used to measure both unrubbed and rubbed fiber contents for 84 peat samples from northern Minnesota. The rubbed fiber contents ranged from 16 to 93% and were on the average 9% lower than the unrubbed fiber content.

When water retention values were related to rubbed fiber content, similar curvilinear relationships were found as was the case with unrubbed fiber contents. The coefficients of determination (R^2) were not significantly different (Table 2). It would appear therefore that either of the values are well-coordinated with physical properties. However, no estimate of actual field unrubbed fiber content was made and it is therefore impossible to know how these values would have correlated with physical properties.

As noted above, the minimum fiber size used in these studies was 0.1 mm. At the time these studies were begun, this was the minimum fiber size being used for classification of peat materials (Farnham and Finney, 1965). The minimum fiber size now used is 0.15 mm.[5] Since a significant relationship also was found between water retention and content of fibers > 0.25 mm, it is logical to assume that a similar relationship would result for a minimum fiber size of 0.15 mm. This is further supported by data showing that fiber contents with minimum fiber sizes ranging from 0.1 to 2.0 mm were equally well-related to bulk density (Boelter, 1969).

It is obvious that the physical properties of peat material are related to

[3]Warren C. Lynn and William E. McKinzie, 8 February 1971. Field tests for organic soils materials; prepared for the 1971 Report of the National Committee on Histosols of the National Technical Soil Survey Work-Planning Conference.

[4]R. S. Farnham, J. L. Bunn, and H. R. Finney, February 1970. Some laboratory methods for analyzing organic soils. Unpublished report, Univ. of Minn., Dep. Soil Science, St. Paul.

[5]Soil Survey Staff, September 1968. Supplement to the soil classification system (7th approximation): Histosols. Soil Conservation Service, USDA. p. 4-1 through 4-7.

Table 2. Curvilinear regression equations and coefficients of multiple determination (R^2) for the relationship of water content (Y) at saturation, 5-mbar, 0.1-bar, and 15-bar suctions to fiber content (X).

Independent variable (fibric content) (X)	Regression equation	R^2
	Saturation	
Unrubbed (>0.1 mm)	$\hat{Y} = 84.36 - 0.039\ X + 0.0021\ X^2$	0.75
Rubbed (>0.1 mm)	$\hat{Y} = 83.54 + 0.046\ X + 0.0015\ X^2$	0.71
Rubbed (>0.25 mm)	$\hat{Y} = 83.97 + 0.086\ X + 0.0012\ X^2$	0.74
	5 mbar	
Unrubbed (>0.1 mm)	$\hat{Y} = 55.22 + 1.478\ X - 0.0169\ X^2$	0.73
Rubbed (>0.1 mm)	$\hat{Y} = 66.42 + 1.146\ X = 0.0160\ X^2$	0.76
Rubbed (>0.25 mm)	$\hat{Y} = 75.44 + 0.814\ X - 0.0146\ X^2$	0.79
	0.1 bar	
Unrubbed (>0.1 mm)	$\hat{Y} = 70.29 + 0.346\ X - 0.0105\ X^2$	0.82
Rubbed (>0.1 mm)	$\hat{Y} = 79.40 - 0.213\ X - 0.0064\ X^2$	0.80
Rubbed (>0.25 mm)	$\hat{Y} = 80.15 - 0.5425\ X - 0.0035\ X^2$	0.83
	15 bar	
Unrubbed (>0.1 mm)	$\hat{Y} = 30.96 - 0.405\ X + 0.0012\ X^2$	0.74
Rubbed (>0.1 mm)	$\hat{Y} = 27.33 - 0.395\ X + 0.0014\ X^2$	0.66
Rubbed (>0.25 mm)	$\hat{Y} = 25.12 - 0.4392\ X + 0.0022\ X^2$	0.70

the degree of decomposition. Frazier and Lee (1971) have also concluded that fiber content is the most useful morphological criterion in classifying organic soils, though the accuracy of determination leaves something to be desired. Because the newly proposed classification of organic soil distinguishes peat material on decomposition as measured by bulk density and fiber content, it should provide useful information about the hydrologic characteristics of the soil.

HYDROLOGY OF ORGANIC SOILS

It has been suggested that peat deposits in Lake States watersheds are streamflow regulators, storing excess snowmelt and rain water in the spring and gradually releasing it during the summer, thus maintaining streamflow. However, based on what we now know about the physical properties of the peat materials, it appears that a more conservative role is played by these organic soil watersheds. Hydrologic data confirm this to be the case.

Organic Soil Profiles

A number of different organic soil profiles exist in Lake States bogs, and horizons of distinctly different peats can be found in a single profile. In the profiles commonly found in undrained lake-filled bogs, the top horizons are undecomposed, forming from the current mosses and woody vegetation. These horizons range from a few centimeters thick in the hollows to 25 to 30 cm in the hummocks. They are in contrast to accumulations of partially decomposed herbaceous peat and occasional thin horizons of more decomposed sapric materials found below the surface.

Different profiles may develop in built-up type bogs found in former glacial lake basins, with the sequence of horizons reflecting the conditions during various stages of development. Some profiles may consist of an extensive horizon of undecomposed sphagnum or moderately decomposed woody or herbaceous peats. The most undecomposed and porous materials are again at the top of the profile, but they may be considerably deeper than in the case of lake-filled profiles.

These less decomposed surface horizons have been termed the *active zone* by the Russian peatland scientists (Romanov, 1961). It is within these horizons that the water table fluctuates and water movement occurs providing zones of drainage and water storage capacity. Once the water table drops into the deeper, more decomposed horizons, little water is removed by drainage.

Hydrogeology of Bog Watersheds

The hydrogeologic situation is a significant factor influencing the hydrology of entire watersheds. Two hydrogeologic conditions have been recognized in the smaller lake-filled bogs of northern Minnesota (Bay, 1966). First is the so-called perched bog with the bog and its water system effectively separated from the regional ground-water system by a layer of fine textured, slowly permeable glacial till. The only source of water to this type of bog is precipitation and some snowmelt runoff in spring of the year. This is in contrast with the ground-water type bog where the water level in the bog is a continuation of the ground-water system. Here the ground-water basin supplies water to maintain water levels in the bog and outflow from the watershed.

Extensive bogs often covering thousands of acres, which have developed in the basins of former glacial lakes, represent an even different hydrogeologic situation. They do, however, have some hydrologic characteristics analogous to the perched bog situation.

Seasonal Distribution of Streamflow

Organic soils, because of their high water storage capacity and flat topography, might be expected to regulate the distribution of streamflow from a peatland watershed. Bay (1969), however, reported this was not the case with small perched, lake-filled bogs. As much as two-thirds of the streamflow from these bogs occurs in the spring, prior to June (Table 3). They ex-

Table 3. Percent of annual water yield by seasons (Bay, 1969).

Watershed and season	Year					5-year average
	1962	1963	1964	1965	1966	
S-1 Spring	70	43	53	43	70	60
Summer	25	56	23	20	26	27
Fall	5	1	24	35	4	13
S-2 Spring	68	45	52	48	67	57
Summer	26	49	22	18	30	27
Fall	6	5	25	32	3	14
S-4 Spring	84	40	69	55	81	69
Summer	15	58	22	15	18	22
Fall	1	1	9	26	1	8
S-5 Spring	84	34	68	50	81	68
Summer	15	65	23	18	17	23
Fall	1	1	9	30	2	9

hibit flow duration curves with steep slopes because streamflow quickly deletes the upper horizons which have very little perennial storage (Figure 3).[6]

High spring runoff from perched bogs results from melting snowpack and early spring rains. Even though the greatest amount of precipitation occurs during late spring, summer, and early fall months, runoff is quite low at these times. Water tables are below the more porous undecomposed horizons and the deeper peats yield too little water too slowly to maintain streamflow. Rainfall is quickly channeled back to the atmosphere by the process of evapotranspiration at the expense of runoff and deep seepage.

While these data show poor distribution of runoff from perched bogs, it appears this also applies to larger built-up bog areas where the primary source of water is precipitation. Winter et al. (1967) reports that a watershed containing a great deal of built-up peatland has a similar distribution of streamflow with low flows characteristic of dry periods. Vidal (1960) reported similar results for organic soil areas in Germany.

The exception to the above is the ground-water type bog which receives much of its water as baseflow from the surrounding ground-water basin. Water levels are maintained at a high level, within the porous, undecomposed horizons and streamflow from the bog is uniformly distributed throughout the year. The flow duration curve has very little slope, showing a nearly constant rate of flow 70% of the time because of the large amount of perennial storage in the basin (Figure 3).

Thus the distribution of flow from organic soil areas is probably not significantly different from mineral watersheds. The hydrogeologic situation is a far more important factor than the organic soil in determining the distribution.

[6]Elon S. Verry, 1970. Water quantity and quality differs greatly between perched and groundwater bogs. Presented at the Lake Superior Biological Conf., Superior, Wis. 25-27 Sept. 1970.

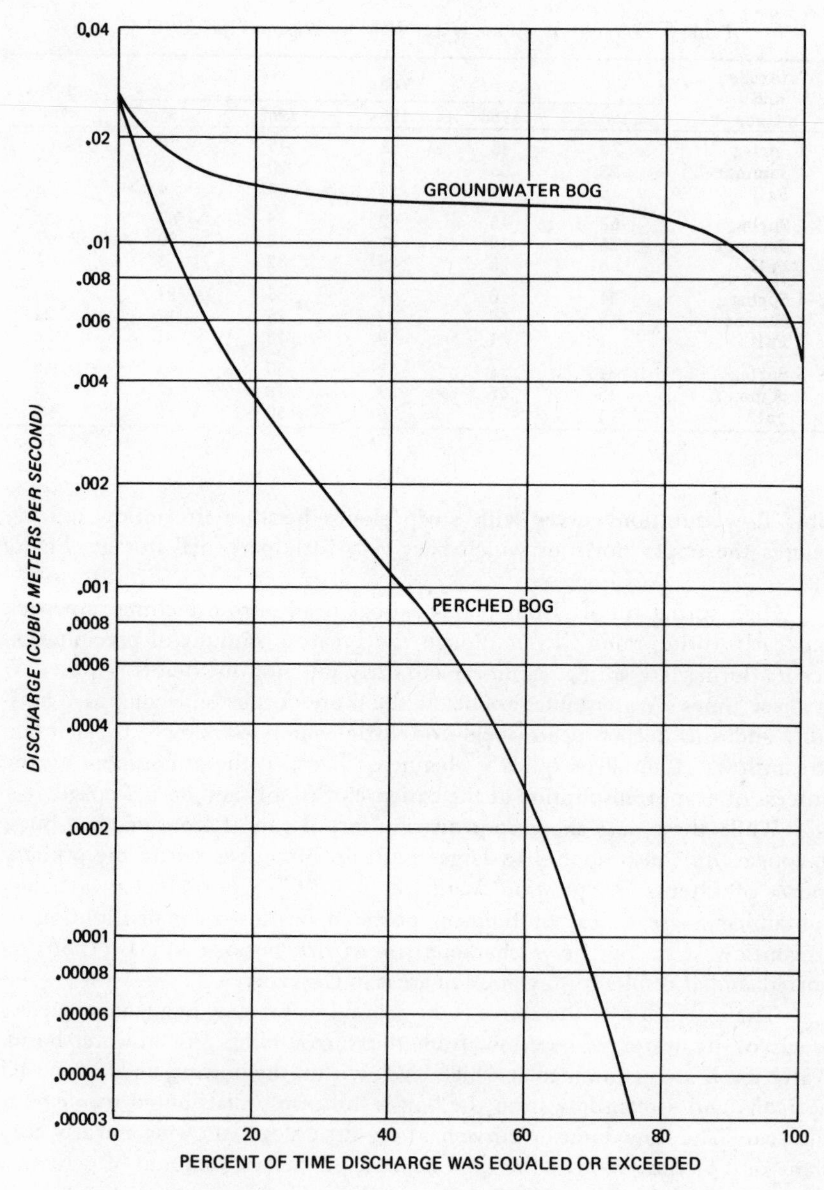

Figure 3. Flow duration curves for perched and ground-water bog watersheds of similar size.

Stormflow

Although bogs generally have only minor effects on seasonal flow, typical storm hydrographs (Figure 4) indicate a temporary storage and slow release

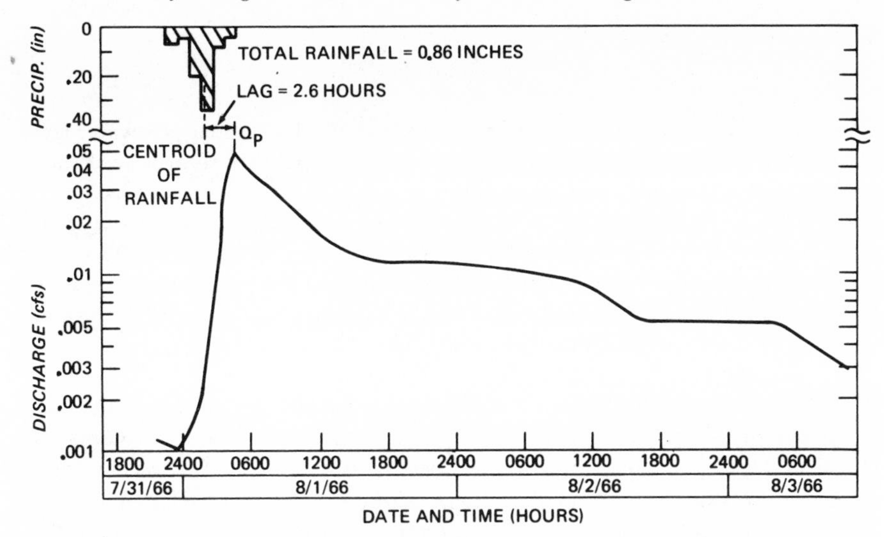

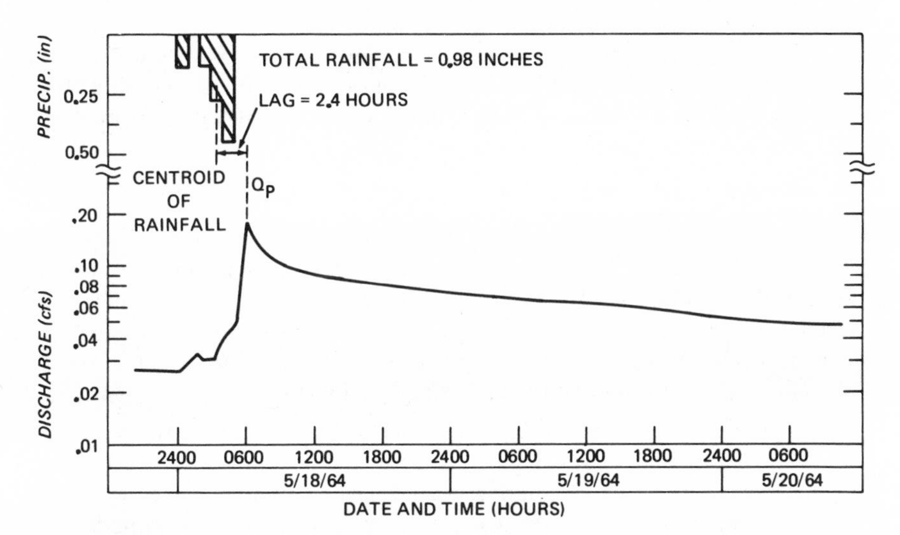

Figure 4. Hydrographs and accompanying hyetographs for two storms from a perched bog watershed (Bay, 1969).

of stormflows due primarily to the near level bog topography and the detention storage of surface peats. The storage capacity is greatly dependent on the position of the water table in the peat profile. Greatest runoff occurs when water tables are high because there is little available storage capacity and water moves directly to the bog outlets. Surface peats also have high hydraulic conductivities and drain quickly while deeper peats are more decomposed, retain more water, and drain very slowly.

Drainage

Because of the natural wet conditions of organic soils, drainage or water level control is a potential method of increasing their capability to produce timber or agricultural crops. Not all attempts at water level management have been successful. The effectiveness varies with the hydraulic conductivity of the peat materials through which the water must flow. Van Bavel (1950) concluded that soils with hydraulic conductivities less than 0.29 feet/day (approx. 1.0×10^{-4} cm/sec) could not be drained economically. Values for peat material are often lower than this (Table 1).

Boelter (1972) reported that an open ditch hastened the drainage of water from upper horizons of undecomposed peat. Once the water table dropped sufficiently so that water had to flow through the more decomposed peat materials to reach the drain, an open ditch had little effect on the water table more than 5 m from the ditch. Burke (1968) reported that a 4.4-m (12-foot) ditch spacing would be necessary to effectively control the water level in blanket peats in Ireland.

If, however, undecomposed peats are quite deep, effective water level control is possible. An open ditch was found to affect the water table 50 m from the ditch in an organic soil with undecomposed peat 1 m or more deep.

SUMMARY

The reader should be reminded that this chapter deals with organic soils in their natural undrained condition. Long-term drainage of organic soils could alter both the physical characteristics of the peat materials and hydrologic characteristics of the organic soils. In the case of undrained organic soils in the Lake States we can draw the following conclusions:

1) Although all peat materials have a high total porosity, the nature of this porosity varies depending upon the decomposition of the material. The least decomposed peats have many large pores which are easily drained and permit rapid water movement. The more decomposed peats have finer pores which hold their water much more tenaciously and permit only slow water movement.
2) Fiber content and bulk density both are related to degree of decomposition and are in turn related to the physical characteristics of the organic soil material. As such they are good criteria for the classification of peat materials.
3) Organic soils are not regulators of streamflow as once was thought. Most runoff from peatland areas occurs in the spring of the year. Once the water table has dropped to the more decomposed peat materials, the peats give up too little water too slowly to maintain streamflow. The exception is ground-water bogs, where due to the base flow of water from the ground-water basin, water levels are

maintained at a high level and the flow is quite uniform throughout the year.

4) Flat topography and the retention storage of peat materials can result in slow release of stormflows on a short term basis. The amount of detention storage will vary greatly depending upon the position of the water table in the peat profile. When the water table is high there is little detention storage capacity and water moves more directly to the bog outlet.

5) The drainage of organic soils is not uniformly successful. In the case of more decomposed materials with low hydraulic conductivity, very close spacing is necessary to effectively control the water table.

LITERATURE CITED

Bay, Roger R. 1966. Factors influencing soil-moisture relationships in undrained, forested bogs. p. 335-343. *In* William E. Sopper and Howard W. Lull (ed.) International Symposium on Forest Hydrology. Pergamon Press, Oxford.

Bay, Roger R. 1969. Runoff from small peatland watersheds. J. Hydrol. 9:90-102.

Boelter, D. H. 1965. Hydraulic conductivities of peats. Soil Sci. 100:227-231.

Boelter, D. H. 1969. Physical properties of peats as related to degree of decomposition. Soil Sci. Soc. Amer. Proc. 33:606-609.

Boelter, D. H. 1970. Important physical properties of peat materials. Int. Peat Congr., Proc. 3rd (Quebec) 1968. p. 150-154.

Boelter, D. H. 1972. Water table drawdown around an open ditch in organic soils. J. Hydrol. 15:329-340.

Baden, W., and R. Eggelsmann. 1963. Zur Durchässignkeit der Moorboden. (Eng. Summary) Zeitschrift Lür Kulturtechnik and Flurbereiningung. 4:226-254.

Burke, W. 1968. Drainage of blanket peat in Glenamoy. Int. Peat Congr., Trans. 2nd (Leningrad, USSR) 1963, II:809-817. Her Majesty's Stationery Office, Edinburgh.

Colley, B. E. 1950. Construction of highways over peat and muck areas. Amer. Highways 29:3-6.

Davis, J. R., and R. E. Lucas. 1959. Organic soils, their formation, distribution, utilization, and management. Mich. State Univ. Agr. Exp. Sta. Spec. Bull. 425. 156 p.

Farnham, R. S., and H. R. Finney. 1965. Classification and properties of organic soils. Advan. Agron. 7:115-162.

Feustel, Irvin C. 1938. The nature and use of organic amendments. USDA Yearbook 1938. p. 462-468.

Feustel, Irvin C., and Horace G. Byers. 1930. The physical and chemical characteristics of certain American peat profiles. USDA Tech. Bull. 214. 25 p.

Frazier, B. E., and G. B. Lee. 1971. Characteristics and classification of three Wisconsin histosols. Soil Sci. Soc. Amer. Proc. 35:776-780.

Hanrahan, E. T. 1954. An investigation of some physical properties of peat. Geotechnique 4:108-123.

Heikurainen, L. 1963. On using groundwater table fluctuations for measuring evapotranspiration. Acta. Forestalia Fennica 76 5:5-16.

Kaila, A. 1956. Determination of the degree of humification of peat samples. J. Sci. Agr. Soc. Finland 28:18-35.

Kuntze, H. 1965. Physikalische Untersuchungsmethoden Für Moorund Anmoorboden. (Eng. summary) Landwirtschaftlishe Forschung 18:178-191.

MacFarlane, Ivan C. 1959. A review of the engineering characteristics of peat. J. Soil Mechanics and Foundations Div., Proc. Amer. Soc. Civil Engineers 85 (SM 1):21-35.

Okruszko, Henryk. 1960. Muck soils of valley peat bogs and their chemical and physical properties. Roczniki Nauk Rolniczych. (Polish translation TT 67-56119, 79 p., 1969).

Romanov, V. V. 1961. Hydrophysics of bogs. Gidrometeorologichejkoe izdatel'stvo. (Russian translation TT 67-51289, 299 p., 1968.).

Van Bavel, C. H. M. 1950. Will this soil drain? Crops and Soils 2(7):10-11.

Vidal, H. 1960. Vergleichende Wasserhaushalts-und Klimabeobachtungen aug unkultivierten und kultivierten Hockmooren in Südbayern. Mitteilungen fur Landkulture, Moorund Torfwirtschaft 8:50-107.

Vorob'ev, P. K. 1963. Investigations of water yield of low lying swamps in western Siberia. Soviet Hydrology: Selected Papers 1963. 3:226-252.

Winter, T. C., R. W. Maclay, and G. M. Pike. 1967. Water resources of the Roseau River Watershed, northwestern Minnesota. Hydrol. Invest. Atlas HA-Z41, Dep. Interior, US Geol. Survey.

Macromorphology and Micromorphology of a Wisconsin Saprist[1]

5

GERHARD B. LEE and BAMRUNG MANOCH[2]

ABSTRACT

Few detailed studies of Histosol morphology were made prior to World War II. Since that time significant investigations have been made in several countries. As a result we know more about their genesis, character, and behavior. The present paper attempts to show how micro- and macromorphological studies, along with other lines of evidence can contribute further to our understanding of Histosols and to their classification. Results of morphological studies, along with chemical and physical analysis, were used to characterize and classify a Wisconsin Saprist. Converging lines of evidence obtained in this manner, provided additional information on the nature of histic materials present, and their structure and fabric.

INTRODUCTION

Histosols (peat and muck soils) comprise a significant part of the natural resource base in several states in this country, and in many other parts of the world. In many places there is intense competition for the use of Histosols, by a wide variety of users.

Because of the importance of these soils, we need to know more about their genesis, and their nature and properties, so that we will be able to predict what kinds of changes occur when they are disturbed. We are concerned here with a group of soils that are in a precarious equilibrium with their environment. Any major disturbance, such as lowering of the water table, results in a series of changes that greatly alters their character and behaviour. Frazier and Lee (1971) pointed out that while fiber content, solubility in sodium pyrophosphate, and several other chemical parameters are extremely useful in the characterization and classification of Histosols, results of these tests did not adequately define all soils presently classified as Saprists and Hemists, and for this reason morphological parameters such as structure and

[1]Research supported by the College of Agricultural and Life Sciences, University of Wisconsin, Madison, and by USAID.

[2]Associate Professor, Department of Soil Science, University of Wisconsin, Madison, and former Research Assistant, now Soil Scientist, Soil Survey Division, Land Development Department, Thailand, resepctively.

fabric might well be used to a greater extent in our classification systems. In addition, micromorphological studies help us understand the nature of the pedogenic processes initiated when raw peaty deposits, accumulated in an anaerobic environment, become aerobic, and are exposed to atmospheric forces, and the concomittant changes in properties and behaviour.

The purpose of this paper, therefore, is to review briefly some of the morphological studies of Histosols to date, and to present results of some recent work on Saprist morphology in Wisconsin.

REVIEW OF LITERATURE

Relatively few detailed morphological studies of histic materials, and Histosols in general, have been made. Among the early studies are those by Dachnowski (1919, 1926) and Dachnowski-Stokes (1933, 1940). In the latter publication the author indicated that "the morphological study of peat areas is a relatively recent line of investigation." Stressing the point that one of the most distinctive features of a peat deposit was its stratification, resulting from accumulation of plant remains under changing environmental conditions, the author suggested that the examination and comparison of different layers in a peat soil led to an understanding of "the epitomized history of external or environmental conditions expressed in positive terms." The character and habits of the peat-forming vegetation, the influences of climate and topography, and the modifying conditions of the ground water were listed as factors instrumental in determining the character of a peat profile.

Somewhat earlier (1926), Dachnowski had described fundamental types of peat profiles and peat materials in terms of their biogenic origin. Included in the latter were reed and sedge peat, hypnum moss peat, sphagnum moss peat, and woody peat.

Soil scientists in Holland (Pons, 1960; van Heuveln, Jongerius & Pons, 1960; Jongerius & Pons, 1962) described peat formation as a geogenetic process in which the parent materials of organic soils are being accumulated. They contrasted this process with the pedologic processes of "ripening" (soil formation) which were initiated by drainage and aeration of a peat deposit. Ripening was described as involving both physical disintegration of plant parts and their biochemical decomposition (moulding). The latter, according to these investigators caused the formation of a "distinct" surface horizon as peat and other material was repeatedly ingested and excreted by soil fauna. Abundant nutrients, low acidity, adequate moisture, and aerobic conditions were noted as environmental factors that tended to encourage faunal activity and accelerate the moulding process. Two kinds of moulded horizons were recognized.

One of these, the "moder" horizon, was described as consisting mostly of fecal excrement from soil fauna such as mites (Collenbala), Diptera, and white pot worms (Enchytralidae). Moder formation occurred in oligotrophic peats containing very little clay, having a pH of 5 or higher, and a carbon/

nitrogen ratio greater than 17. This process did not, however, involve the intimate binding of organic particles necessary to form inseparable humus-mineral complexes as is the case in mull formation. Jongerius (1957) recognized two kinds of moder, namely a small variety 25–60 μm in diameter (after Collenbala, Diptera), and large moder 150–600 μm in diameter (after Enchytralidae). Large and small moder, together with fragments of plant tissue and organic colloids sometimes formed large, loosely aggregated granules called "mull-like moder" by the Dutch workers.

The second moulded horizon called "mull", was described as consisting mainly of earthworm excrement approximately 2 mm in diameter (after Enchytrae, and possibly Julidae). Mull formation was found to occur most commonly under aerobic conditions in eutrophic or mesotrophic peats which contained some clay and were near neutral in reaction. Mull was characterized by a carbon/nitrogen ratio of less than 17. The size and shape of mull aggregates could be altered by a change in environment, for example, continued aerobic conditions caused mull aggregates to coalesce into composites, while prolonged anaerobic conditions apparently caused mull aggregates to disperse into small granules.

Radforth (1952, 1956) described the range of natural structural conditions found in "organic terrain" in Canada. Sixteen structural categories were recognized and illustrated. Radforth and his coworkers have been particularly concerned with the recognition of structural types that can be correlated with stability, bearing capacity, permeability, insulation value and other engineering properties related to trafficability and road construction.

Robertson (1962) discussed the origin and properties of peat, and its use in horticulture. While his primary interest was directed toward proper classification and grading for marketing, he did define several British peat types, in part on the basis of morphology.

Kuiper and Slager (1963) described the occurrence of distinct prismatic and platy structures in Dutch Histosols. They correlated structure formation with a deep ground-water table.

Boelter (1964, 1965) studied the water storage characteristics and hydraulic conductivity of several peat types. His results indicate that these properties vary considerably according to pore size distribution and bulk density of histic materials, as influenced by fiber content and stage of decomposition.

Mackenzie and Dawson (1961) outlined a procedure for thin section preparation and made preliminary studies of structure and fabric in Histosols. Dolman and Buol (1968), in a study of organic soils on the lower coastal plain of North Carolina made note of structure in these soils. Langton and Lee (1964, 1965) studied the structure and fabric of well-decomposed granular horizons in Wisconsin Histosols and outlined a method of thin section preparation by which such horizons might be studied in detail.

In addition, investigators of related soils, or of soil flora and fauna have contributed to our overall knowledge of the morphology of Histosols. Of particular interest is the work of Beryl C. Barratt in New Zealand, who has studied the micromorphology of humus forms and changes in soil structure under various uses (1964, 1967, 1970, 1971), and has published a classification of microscopic soil materials (1969).

RECENT STUDIES IN WISCONSIN

Macromorphology

Field and laboratory studies of Histosols in Wisconsin have been carried out primarily in conjunction with the Soil Survey Program and the North Central Organic Soils Committee.

One fact that is immediately apparent, when looking at cultivated Saprists in southern Wisconsin, is that many of these soils possess distinct pedogenic structure. Comparison of Histosols that have been drained and cultivated for some time, with those in a nearby undrained marsh, or even an undrained portion of the same marsh in which the cultivated soil occurs, show that undrained Histosols do not possess many of the Pedogenic attributes of those that have been drained. Numerous observations of this nature have led to the conclusion that those soils we call Saprists and Hemists in Wisconsin, were formed in fibrous histic materials derived mainly from sedges and rushes, following a lowering of the water table and initiation of an aerobic environment. This may have occurred naturally due to climatic change, stream capture, or the destruction of a beaver dam, or by artificial drainage initiated by man. Following drainage, shrinkage, and rapid initial subsidence, disintegration and decomposition were initiated, resulting in transformation of materials, transfers, and losses, somewhat akin to soil-forming processes in a mineral soil. Additions may have also occurred in the form of sediments from flood waters, eolian dusts, and nutrient elements from fertilizers.

An important element appears to have been the introduction of new forms of soil fauna, e.g., earthworms. In Histosols formed mainly from sedges, rushes, and herbaceous aquatics, morphological indications of pedogenesis include loss of identity of plant materials, reduction in fiber content, destruction of depositional structure and eventually, development of soil structure. Surface horizons become granular or subangular blocky in structure; subsurface horizons become prismatic, with secondary blocky characteristics. The importance of soil fauna as a soil-forming factor is indicated by the presence of biogenic casts or granules throughout the active zone of pedogenesis, particularly in the surface horizon but also in prismatic and blocky peds in the subsoil layers. Pedogenic structure ends abruptly at a level where the soil becomes saturated, and where histic materials, unless affected by an earlier lowering of the water table, retain their fibrous character although they may be partially decomposed.

On a microscale these changes can be observed in much greater detail and further insights gained as to the processes involved and the character of the resultant materials. The following descriptions and other data refer to a

Saprist in University Marsh, Madison, Wisconsin. The original character of this soil, and its history of drainage and use, has been documented in some detail by Huels (1915) and Elliot, Jones, and Zeasman (1921). Morphological, chemical and physical parameters have been studied by Frazier and Lee (1971), J. E. Langton[3] and B. Manoch.[4]

University Marsh, until recently, consisted of approximately 53 ha (130 acres) located in Sec. 16, T7N, R9E, Dane County, Wisconsin along the southwest shore of Lake Mendota. Average annual precipitation is 76.8 cm (30.2 in); the average annual temperature is 7.8C (46.1F).

This area was described by Elliot et al. (1921) as a lake level marsh consisting of true alkaline peat whose surface rose and fell with the level of Lake Mendota. The peat itself was 1 to 6 feet (30 to 180 cm) deep over marl, silts, and clays, and an underlying sand aquifer. Tiling and conversion to cropland began in 1910. Pumping began in 1914. At the end of 5 years total subsidence was 0.76 feet (23 cm) and occurring at a rate of about 0.5 inch/year (1.3 cm) but decreasing.

Native vegetation on the marsh (Huels, 1915) consisted of sedges (*Carex* sp.), bulrushes (*Scirpus*), arrowhead (*Sagittaria*), marsh bluebill (*Campanula*), "marsh grasses," and scattered sphagnum. Wood fragments found in the

[3]J. E. Langton. 1964. Macro- and micromorphology of organic soils. M.S. Thesis, Univeristy of Wisconsin, Madison.

[4]B. Manoch. 1970. Micromorphology of a Saprist. M.S. Thesis, University of Wisconsin, Madison.

Figure 1. Sampling site of Saprist in University Bay Marsh, Madison, Wisconsin.

lower parts of the soil indicated tree growth at one time. Corn was grown on the marsh for 51 years. This marsh is a prime example of changing land use as most of it was covered by fill shortly after it was sampled. It is now used for playing fields and a parking lot. A small area has been retained as a wetland.

Following its original drainage and during the period of cultivation a well-developed Saprist formed from the original peaty deposit (Figure 1). The macromorphology of this soil was as follows:

PROFILE DESCRIPTION; TYPIC MEDISAPRIST, HORICON (HOUGHTON) MUCK

Oal	0–20 cm, black (5YR–10YR 2/1 to N 2/0) muck; ≈ 5% unrubbed fiber; weak, medium, subangular blocky macrostructure; blocks part into moderate, medium and fine granules; soft and very friable moist; many fine roots; many earthworm casts; slightly wavy, clear boundary.
Oa2	20–25 cm, black (5YR 2/1), and dark reddish brown (5YR 2/2) muck; low fiber content; moderate very coarse prismatic structure; prisms part into moderate coarse plates; platy peds coated dark reddish brown (5YR 2/2); many, dark reddish brown (2.5YR 3/4), mainly vertical root tracks; clear wavy boundary.
Oa3	25–36 cm, black (5YR 2/1), and dark reddish brown (5YR 2/2) muck; low fiber content; moderate, very coarse and medium prismatic structure; prisms wedge shaped; they part into weak medium blocks; apparent ped coatings; firm; desication cracks between large prisms ranging from 2 to 20 mm in width. Surface soil noted in some cracks; clear wavy boundary.
Oa4	36–41 cm, black (5YR 2/1), and dark reddish brown (5YR 2/2) muck; low fiber content; weak to moderate, very coarse prismatic structure; prisms part into moderate coarse plates; reddish brown (2.5YR 4/3) root tracks; abrupt, wavy boundary.
Oa5	41–61 cm, black (5YR 2/1), and reddish brown (2.5YR 3/4) finely divided, dense muck; ≈ 10% fiber; moderate, medium and fine, prismatic macrostructure; both horizontal and vertical prism faces appear to be coated; many earthworm casts in voids between large prisms; clear wavy boundary.
Oa6	61–81 cm, black (5YR 2/1), muck; weak coarse prismatic macrostructure; thin, glossy, continuous coatings on prism faces; many vertical roots; very moist; clear boundary.
Oa7	81–99 cm, dark reddish brown (5YR 2/2) peaty muck; becomes dark gray on exposure to air; weak coarse prismatic structure *in situ*; a few woody stems 1 to 2.5 cm diameter; very moist; horizontal cleavage face at lower boundary; very wet at this interface; clear boundary.
Oe1	99–117 cm, brown to dark brown (7.5YR 4/4) mucky peat; be-

comes dark grayish brown (10YR 3/2) on exposure to air; many fine fibers; matted; spongy, and saturated; clear boundary.

Oe2 117–142 cm, reddish brown (5YR 4/4) peat; becomes very dark grayish brown (10YR 3/2) when exposed to air; ≈ 20% fiber, mainly fine; woody stems common; clear boundary.

Oe3 142–142 cm, brown to dark brown (7.5YR 4/4) peat; ≈ 30% fiber, coarser than above; appears to be derived from sedges; massive, saturated; identifiable sedge remains; clear boundary.

Oe4 152–183 cm, dark yellowish brown (10YR 3/4) peat; becomes gray (10YR 5/1) with exposure to air; estimated 30% fiber; massive, dense, compact; many seed pods and wood stems, clear boundary.

Oi1 183–193 cm, a layer of coarse fibrous material at boundary with underlying sedimentary material; abrupt, smooth boundary.

II Lco and Lca 193+ cm, very dark gray (10YR 4/1) sedimentary peat and marl; massive.

As can be seen from this description, and the accompanying laboratory data (Table 1), this soil bears little resemblance to the original peaty deposit described by Elliot et al. (1921). Pedogenesis over a period of about 60 years, since drainage was first initiated and cultivation began, had resulted in destruction of most fiber in the upper 100 cm (Table 1). Biochemical decomposition had also taken place as exhibited by dark, essentially black colors, narrow C/N ratios, and brown sodium pyrophosphate extract colors (SPEC). Pedogenic structure was also evident, expressed in the granularity of the surface horizon, and the prismatic/blocky expression of subsurface layers (Figure 2). The secondary platy structure of several subsurface layers, as observed *in situ*, appeared to be a relict expression of the original depositional structure of the parent vegetative deposits.

Table 1. Physical and chemical characteristics of University Marsh Saprist (data from Frazier & Lee, 1971).

Horizon or layer	Depth	Fiber*	SPEC†	pH	Ash	C‡	N‡	C/N
	cm	%	10YR			%		
Oa1	0–20	3	5/4	5.7	26.2	55.8	4.3	13
Oa2	20–25	ND	6/3	5.7	13.4	55.8	3.7	15
Oa3	25–36	7	5/3	5.8	24.5	57.7	4.2	14
Oa4	36–41	ND	5/4	5.9	22.2	57.1	4.0	14
Oa5	46–61	16	6/3	6.1	19.7	57.4	4.1	14
Oa6	61–81	9	6/3	6.2	27.5	58.3	4.2	14
Oa7	81–99	14	6/3	6.0	17.2	58.7	3.2	18
Oe1	99–117	ND	5/4	4.3	14.4	58.7	3.0	20
Oe2	117–142	37	6/4	5.8	15.5	59.7	2.6	23
Oe3	142–152	47	7/2	5.6	11.0	58.6	2.7	22
Oe4	152–183	41	7/1	5.9	10.8	58.3	3.5	17

* Percent by weight of organic fraction retained on 140-mesh (105-μm) screen.
† Acronym for Sodium Pyrophosphate Extract Co. or. Notations are for Munsell color chips. ‡ Percent by weight of organic fraction.

Figure 2. Close-up of Saprist profile showing granular plow layer underlain by prismatic-blocky structured subsoil. Tape shows depth in feet.

Several morphological features of interest are shown in Figures 3 to 5. Among these are the prismatic/blocky structure of subsurface sapric horizons shown in Figure 3. The peds shown are from the Oa5 horizon (41–61 cm) which was described in the field as a dense muck. Fiber content of this layer was 16% by weight of the organic fraction; ash content was approximately 20% of the total weight (Table 1). Figure 4 shows ped surfaces in somewhat greater detail. Apparent coatings can be seen along faunal channels and on portions of the ped surface. Hemic material is identified by its honey comb appearance.

Faunal casts were clearly visible in several horizons but were particularly noticeable in horizon Oa5. Figure 5 shows a cluster of casts on a vertical prism surface.

The structural characteristics of the sapric portion of this soil, the apparent coatings on peds, granular surface material found at depth in voids between prisms, and abundant faunal casts, all appear to be morphological in-

Figure 3. Prism-like ped from Oa5 horizon (41–61 cm). Note secondary blocky structure.

dicators of active pedogenic processes in which biotic processes play a particularly important role. Somewhat similar processes have taken place or are occurring on Alfisols adjacent to the marsh, for example, a high degree of mixing of A1 and A2 horizons can be seen in these soils. While several species of earthworms, mites, and other fauna are likely involved, the most obvious species appears to be *Lumbricus terrestris* which is believed to have been imported accidently from Europe *circa* 1890 (Nielsen & Hole, 1964), and possibly carried to the area by fishermen.

While the character, distribution, and genesis of ped coatings in Histosols remains somewhat conjectural at this time, observations to date suggest that they are partly biotic and partly illuvial in origin, and have a higher mineral content then ped interiors. It is possible that the mineral content of illuvial coatings on prisms and in some channels has been augmented by sediment laden runoff waters from the surrounding slopes, particularly in years past when some of the area was cultivated.

Figure 4. Detail of ped surface showing coatings along faunal channels and on portions of ped surface. Included hemic material has honey combed appearance.

At the time of sampling the soil was unseasonably dry in its upper layers and conspicuous dessication cracks were present both at and below the surface. Such cracks likely provided the means whereby soil material could slough off the very granular and friable surface horizon and fall to lower depths during dry periods. When a dry period was followed by storms of such intensity as to cause runoff the structural cracks allowed water born sediment to flow to lower layers. At depths of 81 to 99 cm the soil was moist when sampled; structure, although present, was only weakly expressed. Below 99 cm the soil was saturated. Field observations indicated that the layers from 99 to 142 cm were more fibrous then the horizons above, and lacked structure, however, a reduction in fiber size from the original plant tissue suggested that physical disintegration had begun. Brown pyrophosphate colors (Table 1) indicated biochemical decomposition as well.

Soil material from 142 to 183 cm was also fibrous in appearance, differing from the material immediately overlying it by having a relatively high

Figure 5. Cluster of faunal casts in void between prisms. Individual casts are on order of 3 to 6 mm diameter.

Figure 6. Prism-like clod from upper part of Limnic horizon. Note thin layer of fibrous peat on surface of layer and snail shells scattered throughout. Scale is approximately 1:1.

Table 2. Composition of limnic layer in University Marsh.*

		%
Mineral matter (sand, silt, clay)		51.3
Organic matter		9.8
Calcite (shells and disseminated $CaCO_3$)		41.3
	Total	102.4
Mineral matter (from above)		%
Sand		36.6
Silt		47.1
Clay		15.1
	Total	98.8

* Unpublished data J. Langton and G. B. Lee.

content of identifiable fibers and other plant remains. This material terminated at a depth of 183 cm, above a 10-cm thick layer of very coarse fibrous material which in turn was underlain by marly sedimentary peat.

The latter material, a limnic deposit, ranged in thickness from a few centimeters to a decimeter or more in various parts of the marsh. Composition was also variable. In some places (mainly shallow areas) it was extremely marly whereas in deeper parts of the bog it had more of the characteristics of coprogenous earth although both disseminated and shelly calcitic material was included as were noncarbonate detrital sediments. Figure 6 shows the gross character of this horizon at the site sampled.

Earlier studies of the composition of this layer, at a nearby site, showed that it consisted of approximately 50% mineral matter and about 40% carbonate by weight (Table 2). The latter consisted mainly of calcite, whereas the clay of the mineral fraction consisted mainly of 18Å chlorite-montmorillonite intergrades (unpublished data, J. Langdon and G. B. Lee).

Micromorphology

Thin sections, from major horizons, impregnated with Carbowax 6000,[5] as described by Langton and Lee (1965), were studied by use of stereoscopic and petrographic microscopes. Classification of materials present followed the scheme outlined by Langton (1964) as modified by the authors of this paper.[6]

Brief micromorphological descriptions of selected horizons from each of four major zones, and interpretations, are as follows:

Oalp, 0 to 20 cm—This has many pelletized granules, including brown moder (0.012–0.90 mm) and dark-colored mull (0.2–1.5 mm), in a matrix of brown amorphous material (Figure 7). Common black fragments are scattered

[5]Union Carbide Chemical Co., New York, N. Y.

[6]Terms used are mainly descriptive. Relative area of slide occupied by constituents noted indicated by Few = 0–2%; Common = 2–20%, Many > 20%.

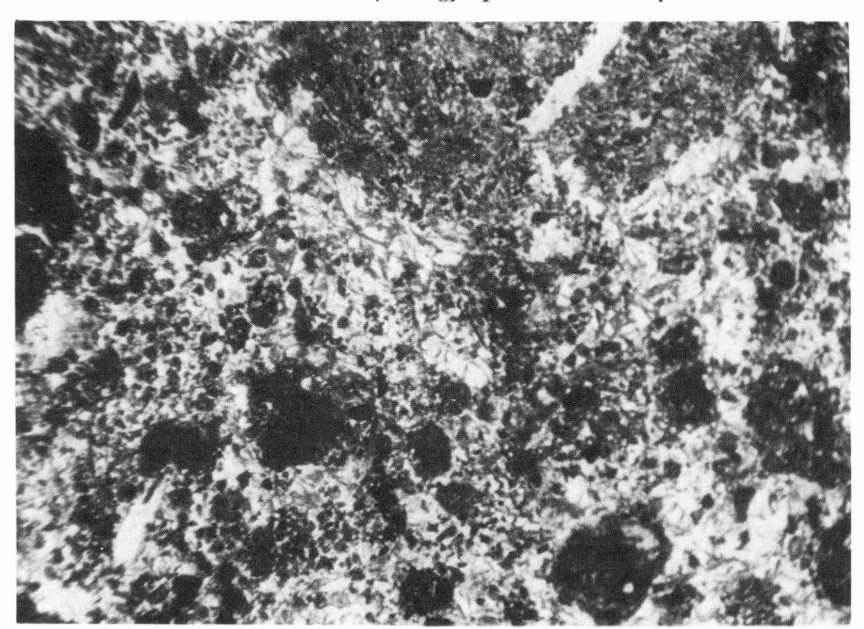

Figure 7. Thin section of Oa1p horizon. A portion of a mull aggregate is shown in the upper part of the section. Moder and black fragments are common. Light areas are voids. (X 25).

throughout the matrix or included in moder and mull aggregates. It contains few, brown, finely disintegrated herbaceous fragments; few pollen grains and faunal skeltal remains.

Analysis (Table 1) of this horizon showed that it was very low in fiber (3%) and well decomposed (SPEC = 5/4). Nitrogen content was 4.3%; the C/N ratio was 13; with pH 5.7. Earlier studies of this horizon (Langton, 1964), showed it to have relatively high bulk density (0.32 g/cc) and relatively low water content when saturated (202% by weight).

This horizon was classified as a well-decomposed (sapric) horizon in the field. Laboratory and micromorphological studies confirmed this classification and revealed in more detail the fabric and nature of materials present.

Oa5, 41 to 61 cm—The primary constituents are many black fragments aggregated into cell-like structures approximately 75 μm in diameter, in a black opaque matrix. Brown amorphous material is present on ped surfaces and in laminae. Included brown hemic material has relict cell structure.

Analysis (Table 1), showed this horizon to be moderately low in fiber (16%) and moderately well decomposed (SPEC = 6/3). Nitrogen content was 4.1%; the C/N ratio was 14; with pH 6.1. According to Langton (1964), bulk density of this zone was 0.25 g/cc; saturated water content was 309% by weight.

Field studies indicated that this horizon consisted of dense muck possessing prismatic structure with coatings on both vertical and horizontal ped faces (Figure 4). Micromorphology suggests that prisms consisted of sapric *and hemic* material coated in part with amorphous illuvial material.

Oe3, 142 to 152 cm—The soil is made of mainly fibrous material with woody stems common; few to common brown and black fragments; brown amorphous coatings noted along root channels; few to common seeds; pollen grains, and sand-size mineral particles.

Analysis (Table 1) of this horizon showed abundant fiber (47%); SPEC was 7/2 indicating little decomposition. Nitrogen content was 2.7%; the C/N ratio was 22; with pH 5.6. Bulk density in this zone (Langton, 1964) was 0.21 and saturated water content 385% by weight.

Field observations indicated that this was a fibric horizon. Laboratory analysis and micromorphological studies confirm its gross composition as noted in the field but also disclosed evidences of decomposition that suggest it should be classified as hemic material.

IILco and Lca, 193 to 213 cm—This is made up of a heterogenous mixture of finely disintegrated plant materials, fecal pellets, shell fragments, seeds, pollen grains, and detrital mineral sediments.

Micromorphological studies suggest that this layer is a complex mixture of materials and contains more mineral detritus than current definitions of marl or coprogenic earth would suggest.

SUMMARY

Results of morphological studies showed that while a variety of both organic and inorganic constituents were present throughout the soil, surface, subsurface, and substratum horizons each had definitive macro- and micromorphological characteristics.

The sapric surface horizon was found to be very granular in structure and low in fiber content; it likely formed under aerobic conditions. On the basis of its high population of faunal aggregates (mor and moder), the good granular structure of this horizon is believed to be related to faunal activity subsequent to artificial drainage and initiation of cultivation of University Bay Marsh some 50 years ago.

Sapric subsurface horizons consisted mainly of black and brown histic fragments in a brown amorphous matrix along with occluded fragments of hemic material. Cell-like micropeds in this horizon were aggregated into blocks and prisms, clearly visible in field studies. Apparent ped coatings, suggestive of illuviation, were noted. Faunal aggregates on ped surfaces (Figure 5) and a worm with fecal pellets in its gut (Figure 8), are noted as positive indications of the active role of soil fauna in the continuing pedogenesis of the subsoil zone.

Hemic layers in the substratum were more fibrous than horizons in surface and subsurface zones, and lacked pedogenic structure. However, micromorphological studies of these underlying materials revealed that they were

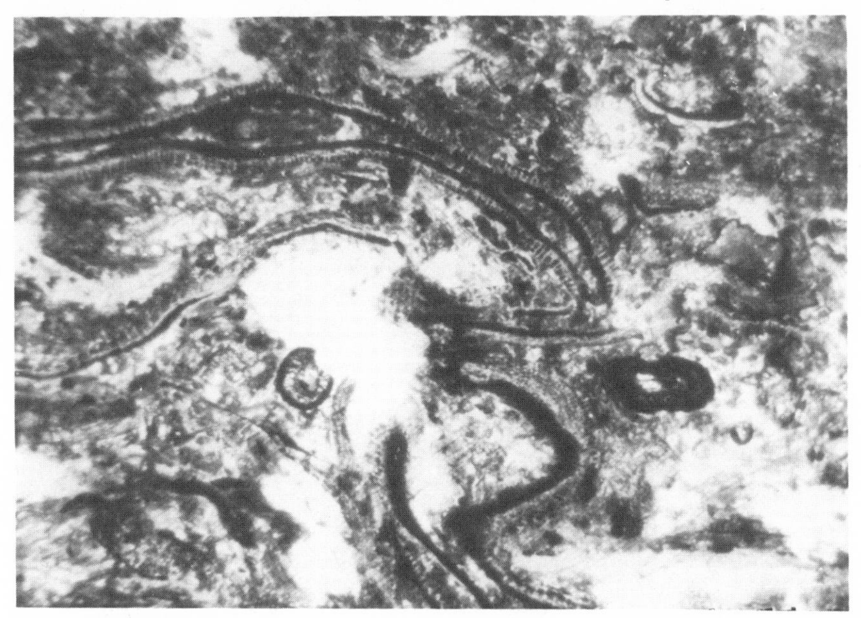

Figure 8. Thin section of Oa7 horizon, just above hemic material (Oe1). Worm in upper foreground with fecal pellet in gut. (X 25).

more decomposed than indicated by field studies and that locally, amorphous materials were present in root channels.

CONCLUSIONS

In general, micromorphological characteristics correlated very well with stage of decomposition and pedogenesis as determined by macromorphological methods and by physical and chemical analysis. In addition, such studies provided additional information on the nature of the material present, its structure and fabric, and its genesis. By this means micromorphology contributes to our overall understanding of Histosols, and improves our ability to make valid predictions and recommendations regarding their use and management.

LITERATURE CITED

Barratt, B. C. 1964. A classification of humus forms and microfabrics in temperate grassland. J. Soil Sci. 15:342–356.

Barratt, B. C. 1967. Differences in humus forms and their microfabrics induced by long-term topdressings in hayfields. Geoderma 1:209–227.

Barratt, B. C. 1969. A revised classification and nomenclature of microscopic soil materials with particular reference to organic components. Geoderma 2:257–271.

Barratt, B. C. 1970. Effect of long-term fertilizer topdressing in hayfields on humus forms and their micromorphology. Agr. Digest 21. New Zeal. Soil Bureau Publication 498. p. 11–18.

Barratt, B. C. 1971. A micromorphological investigation of structural changes in the topsoil of Potumahoe clay loam used for market gardening. New Zeal. J. Sci. 14:580-598.

Boelter, D. H. 1964. Water storage characteristics of several peats *in situ.* Soil Sci. Soc. Amer. Proc. 28:433-435.

Boelter, D. H. 1965. Hydraulic conductivity of peats. Soil Sci. 100:227-231.

Dachnowski, A. P. 1919. Quality and value of important types of peat material. USDA Bull. 802. 40 p.

Dachnowski, A. P. 1926. Factors and problems in the selection of peat lands for different uses. USDA Bull. 1419. 23 p.

Dachnowski-Stokes, A. P. 1933. Grades of peat and muck for soil improvement. USDA Circ. 290.

Dachnowski-Stokes, A. P. 1940. Research in regional peat investigations. J. Amer. Soc. Agron. 22:352-366.

Dolman, J. D., and S. W. Buol. 1968. Organic soils on the lower coastal plain of North Carolina. Soil Sci. Soc. Amer. Proc. 32:414-418.

Elliot, G. R. B., E. R. Jones, and O. R. Zeasman. 1921. Pump drainage of the University of Wisconsin marsh. Univ. of Wis. Agr. Exp. Sta. Res. Bull. 50. 32 p.

Frazier, B. E., and G. B. Lee. 1971. Characteristics and classification of three Wisconsin Histosols. Soil Sci. Soc. Amer. Proc. 35:776-780.

Huels, S. W. 1915. The peat resources of Wisconsin. Wis. Geol. Nat. Hist. Survey Bull. XLV. 274 p.

Jongerius, A. 1957. Morfologishe onderzoekingen over de boemstructure. Bodemkundigl Studies No. 2, Wageningen.

Jongerius, A., and L. J. Pons. 1962. Some micromorphological observations on the process of moulding in peats in The Netherlands. 2 Pfl Ernahr Dung 97:243-255.

Kuiper, F., and S. Slager. 1963. The occurrence of distinct prismatic and platy structures in organic soil profiles. Neth. J. Agr. Sci. 11:418-421.

Langton, J. E., and G. B. Lee. 1964. Characteristics and genesis of some organic soil horizons as determined by morphological studies and chemical analysis. Wis. Acad. Sci. Arts Letters 53:149-157.

Langton, J. E., and G. B. Lee. 1965. Preparation of thin sections from moist organic soil materials. Soil Sci. Soc. Amer. Proc. 29:221-223.

Mackenzie, A. F., and J. E. Dawson. 1961. The preparation and study of thin sections of wet organic soil materials. J. Soil Sci. 12:142-144.

Nielsen, G. A., and F. D. Hole. 1964. Earthworms and the development of coprogenous Al horizons in forest soils of Wisconsin. Soil Sci. Soc. Amer. Proc. 28:426-430.

Pons, L. J. 1960. Soil genesis and classification of reclaimed peat soils in connection with initial soil formation. Int. Congr. Soil Sci. Trans. 7th (Madison, Wis.) 1960. IV:205-211.

Radforth, N. W. 1952. Suggested classification of muskeg for the engineer. Eng. J. 35: 1199-1210.

Radforth, N. W. 1956. Range of structural variation in organic terrain. Tech. Memo No. 39. National Research Council, Ottowa, Canada. 67 p.

Robertson, R. A. 1962-1963. Peat: Its origin, properties, and use in horticulture. Sci. Hort. 16:42-52.

van Heuvelin, B., A. Jongerius, and L. Pons. 1960. Soil formation in organic soils. Int. Congr. Soil Sci. Trans. 7th (Madison, Wis.) 1960. p. IV:195-204.

Physical, Chemical, Elemental, and Oxygen-Containing Functional Group Analysis of Selected Florida Histosols[1]

6

L.W. ZELAZNY and V.W. CARLISLE[2]

ABSTRACT

Four soil series selected for investigation represented the most diverse natural morphological and decompositional ranges of Histosols occurring within the Florida Everglades. The least decomposed Histosol was represented by the Montverde series, with Pahokee representing an intermediate stage of decomposition, and Okeelanta representing the most advanced stage. Although the surface of Torry represented an advanced stage of decomposition, it was also examined because of exceptionally high clay content and significantly lower rate of subsidence. Stages of decomposition decreased with increasing depth within series due to decreasing oxidative conditions. Therefore the least decomposed horizon was represented by the 36-to 81-cm horizon of Montverde followed by the 109 to 137-cm horizon of Torry, and the most decomposed horizons were the surface horizons of Okeelanta and Torry.

Bulk density and percent ash decreased with increasing profile depth, and generally increased with Montverde < Pahokee < Okeelanta < Torry. Generally the water-retaining characteristics increased with increasing profile depth and decreased with Montverde > Pahokee > Okeelanta > Torry. The humin fraction generally constituted the highest organic fraction percentage and increased with profile depth. Comparison of surface horizons show Montverde to have the largest humin fraction percentage, Pahokee and Okeelanta to be intermediate, and Torry to be lowest. Fulvic acid fraction percentage was lowest for the deeper horizons of both Montverde and Torry, and highest for the surface horizons of Pahokee and Okeelanta. The first two horizons of Torry contained the highest clay and humic acid percentage and the largest humic acid/fulvic acid ratios. Total elemental analysis revealed striking similarities for all horizons and organic fractions. Similarities for all horizons also occurred with functional group analysis except for the two clayey horizons of Torry which had a higher carboxyl group content and a lower phenolic hydroxyl group content in the humin and humic acid fractions. Generally, total acidic and carboxyl group content was greatest in the fulvic acid fraction and lowest in the humin fraction. Conversely, the humin fraction contained the largest alcoholic hydroxyl group content.

---

[1]Contribution from the Soil Science Department, University of Florida, Gainesville, Florida.

[2]Assistant Soil Physical Chemist and Mineralogist, and Associate Professor of Soil Science, respectively.

INTRODUCTION

The largest contiguous body of organic soils in the continental United States occurs in the Florida Everglades. It is primarily a great sawgrass marsh (*Cladium jamaciénse* Crantz) which lies in a trough approximately 65 km wide and 160 km in length that extends from Lake Okeechobee to nearly the end of the Florida peninsula (Figure 1). Both sides are bounded by low sandy ridges. Elevations are approximately 5 m above sea level at the south shore of Lake Okeechobee (Cooke, 1945) with a slope from north to south to about 5 cm/km (Clayton, Neller, & Allison, 1942). Depth of organic materials, however, varies greatly from north to south. Near the east side of Lake Okeechobee the depth of organic materials is 2.5 to 3.5 m but in the southern part of the Everglades it is quite shallow (Davis, 1946). The depth over a large part of the area is now < 1 m and probably $< 200{,}000$ ha has a depth > 1.5 m.

Most of the Everglades area is underlain by a more or less porous deposit of limestone and marl known as the Fort Thompson formation (Jones, 1942). It underlies the area adjacent to Lake Okeechobee and extends south for approximately 90 km. West of the Palm Beach County line a layer of sand usually occurs between the organic materials and rock.

The Florida Everglades contain about 75% of the organic soils in the state (Davis, 1946) and about 14% of the total national deposits (Stephens, 1969). Approximately 283,000 ha of the Everglades are suitable for agriculture of which nearly 182,000 ha are currently utilized for crop production. This is one of the richest agricultural regions on earth, providing much of the nation's winter vegetables, in addition to supporting large sugarcane, cattle, and sod enterprises.

Histosols of the Everglades are subsiding at a rate of approximately 3 cm/year (Stephens, 1956; 1969). This amounts to over 135 metric tons of CO_2 released into the atmosphere from each hectare each year (Knipling, Schroder, & Duncan, 1970). Stephens and Johnson (1951) prepared isopachous charts which indicated loss of soil volume in the area studied that amounted to 40% of that present between the first surveys in 1912 and the 1950 survey. They predicted total volume loss in cross-sectional area from predrainage days would be 66% by 1970 and 88% by 2000. These predicted amounts of subsidence are very close to current measurements (Stephens, 1969).

While all causes of subsidence are important, biochemical oxidation is the most harmful and uncontrollable. Biochemical oxidation proceeds with decomposition of the most readily oxidizable compounds to less decomposable compounds with the evolution of CO_2. According to this principle, organic soils should become less oxidizable with time; however, rates of Histosol decomposition in the Everglades have not decreased (Neller, 1944; Stephens, 1969). "With continued subsidence, by 1990 most of the organic soils in the Everglades will be too shallow to support a paying agriculture,

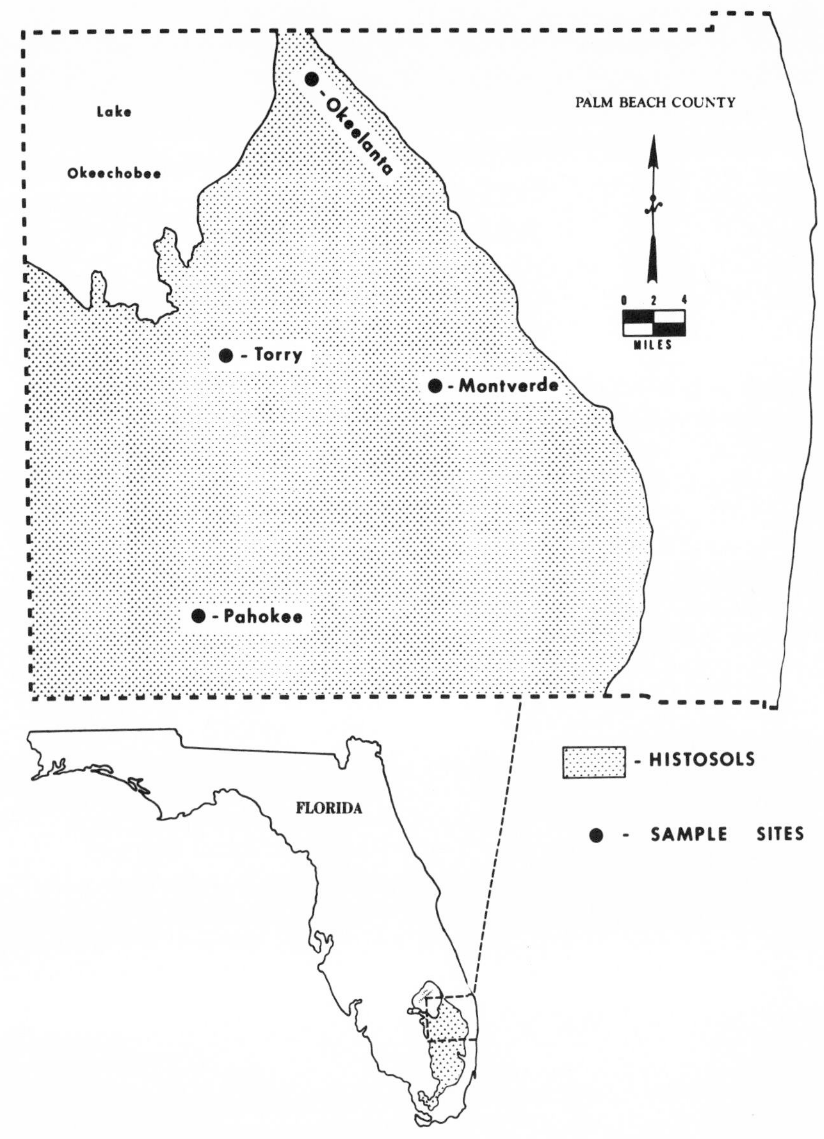

Figure 1. Histosol area of Florida Everglades and location of profile sample sites in Palm Beach County.

and by the turn of the century will have subsided to the point of widescale abandonment." (Stephens, 1969).

Regulation of the water table presently offers the best means of minimizing subsidence. Changes occurring in organic matter with continued biochemical oxidation are unknown, hence subsidence is not fully understood and cannot be controlled in drained areas. The present investigation was undertaken to characterize selected Histosols of the Florida Everglades according to physical and chemical parameters, elemental and oxygen-containing functional group analysis of the humin, humic acid, and fulvic acid fractions, and to determine characteristic differences in the organic constituents of these soils.

MATERIALS AND METHODS

Sampling Procedures

All soils investigated were from sugarcane (*Saccharum officinaram* L.) fields located in Palm Beach County, Florida (Figure 1). Individual sample sites were located with aid of the Soil Conservation Service field Soil Survey Party. Soils selected for investigation represented the widest possible morphological and decompositional ranges of Histosols occurring within the Florida Everglades. At each sample site the freshly exposed profile was carefully examined and described according to procedures outlined in *Soil Taxonomy* (Soil Survey Staff, 1974). Differences in horizon designation and series classification (Table 1) since time of sampling, 1972, reflect recent changing concepts in Histosol taxonomy.

Montverde soils, represented by Profile 1, developed in herbaceous organic deposits underlain by limestone. These soils represent the least decomposed Histosols of the Everglades, occurring only in areas that have organic deposits extending to depths of 130 cm or more. Pahokee soils, represented by Profile 2, developed from the remains of nonwoody, hydrophytic plants underlain by limestone. These soils represent an intermediate stage of decomposition as compared to other Histosols in the Florida Everglades. They represent the most extensive condition prevailing in the Everglades, namely, $<$ 130 cm of organic materials over hard limestone. Okeelanta soils, represented by Profile 3, developed from hydrophytic nonwoody plant remains

Table 1. Comparative pedon taxonomy.

Profile no.	Series	Depth	Horizon designation		Classification	
			March 1972	May 1973	March 1972	May 1973
		cm				
1	Montverde	0-13	Oap	Oap	Typic Medifibrist	Typic Medisaprist
		13-36	Oe	Oa2		
		36-81	Oi	Oa3		
2	Pahokee	0-18	Oap	Oap	Lithic Medihemist	Lithic Medisaprist
		18-86	Oe	Oa2		
3	Okeelanta	0-21	Oap	Oap	Terric Medisaprist	Terric Medisaprist
		21-66	Oa2	Oa2		
4	Torry	0-21	Oap	Oap	Typic Medisaprist	Typic Medisaprist
		21-53	Oa2	Oa2		
		53-109	Oa3	Oa3		
		109-137	Oe	Oa4		

mixed with various amounts of mineral material, mostly sands. These soils represent the most advanced stages of decomposition, occurring along the edges of the Florida Everglades and as narrow elongated areas within the Everglades. Torry soils, represented by Profile 4, developed from hydrophytic nonwoody plant remains mixed with fine-textured mineral materials. These soils, placed in the clastic family grouping because of high clay content, represent advanced stages of decomposition. They occur in a relatively confined area along the southeastern shore of Lake Okeechobee.

An approximate 5-kg soil sample from designated horizons within each profile was collected into labeled plastic bags, put in sealed plastic containers and placed under refrigeration to retard oxidation. One set of triplicate undisturbed core samples (5.4 by 6.0 cm) was obtained from each horizon for bulk density and percent ash determinations. Another set of triplicate core samples (5.4 by 3.0 cm) was obtained from each horizon for determining water retaining and transmitting properties. Core sampling was interspersed throughout each designated soil layer to represent observable variabilities in horizon morphology.

Analytical Procedures

All determinations were performed in triplicate on undried samples. Weights were based on drying separate aliquots in an oven at 110C. Soil reaction was measured in a 1:1 soil-solution volume ratio using both distilled H_2O and 1*N* KCl. Exchangeable bases were extracted with a 1:100 weight ratio soil-neutral 1*N* NH_4OAc. Calcium and magnesium were analyzed by using a Perkin Elmer Model 303 atomic absorption spectrophotometer, and Na and K by using a Beckman B flame spectrophotometer. Soil acidity was determined by extraction with 0.5*N* $BaCl_2$ and excess 0.055*N* triethanolamine adjusted to pH 8 with subsequent titration of unreacted triethanolamine with standard HCl to the end point of bromocresol green–methyl red mixed indicator.

Core samples obtained for calculating bulk density values were oven-dried at 110C for 24 hours. These soil samples were subsequently heated in a muffle furnace at 500 and 700C for 6 hours to obtain percent ash at these two temperatures. The cores obtained for determining water retaining and transmitting properties were placed in Tempe pressure cells, saturated, and sequentially extracted at pressures of 30, 60, 100, 150, 200, and 345 mbars. Water contents were determined from cell weight at each equilibrium pressure. The water-retaining characteristics were plotted and values selected at 0.10 and 0.33 bars. The cores were resaturated for determination of saturated hydraulic conductivity and subsequently transferred to a ceramic plate extractor to determine 15-bar water retention.

Chemical fractionation of organic matter was performed in triplicate on undried samples. Samples were first acid washed with 0.1*N* HCl and once with deionized H_2O to remove HCl. Insoluble humin fractions were separated from the alkali-soluble humic acid and fulvic acid fractions by shaking a 1:100 soil–0.5*N* NaOH solution for 24 hours under N_2. Humin fractions were separated from soluble organic fractions by centrifugation at 2,000 rpm for 15 min in an International No. 2 centrifuge and decantation. The material remaining after decantation was washed three times with fresh 0.5*N* NaOH, acidified with HCl to pH 2, dialized against distilled H_2O until Cl free as determined by $AgNO_3$ test, and then freeze-dried. Acid-insoluble humic acid fractions were separated from acid-soluble fulvic acid fractions by acidifying the NaOH-extract with HCl to pH 2. Fulvic acid fractions were removed by siphoning off the supernatant liquid followed by filtration through Whatman No. 42 filter paper. Humic acid fractions were washed once with 0.01*N* HCl, dialized against distilled H_2O until Cl free as determined by $AgNO_3$ test, and then freeze-dried. Fulvic acid fractions were desalted by passage through a hydrogen-saturated column of Amberlite IR-120 exchange resin and concentrated with a rotary evaporator operating at 50C and reduced pressure. All fractions were stored in vacuum desiccators over P_2O_5 and weighed to determine total yield. Ash contents were determined by heating an aliquot in a muffle furnace to 700C for 6 hours, and weighing. Organic fractions containing $> 5\%$ ash were shaken for 3 hours with 1:200 organic fraction–0.5% HCl-HF solution. Residues were filtered on Whatman No. 42 filter paper, washed with H_2O until Cl free as determined by $AgNO_3$ test, and dried in a vacuum desiccator over P_2O_5. These treatments were continued until the organic fraction contained $< 5\%$ ash. These fractions were subsequently analyzed for chemical components and oxygen-containing functional groups. All results were expressed on a moisture-free and ash-free basis.

Carbon and hydrogen were determined by dry combustion using a Coleman Model 33 Carbon-Hydrogen Analyzer. Nitrogen was analyzed by micro-Kjeldahl technique. Sulfur was analyzed by a dry combustion technique (Tiedemann & Anderson, 1971) with the use of a Leco induction furnace and automatic sulfur titrator. The remaining composition was assumed to be oxygen.

Total acidity was determined by equilibration with excess $Ba(OH)_2$ solution and potentiometric titration of unreacted $Ba(OH)_2$ with standard HCl (Schnitzer & Gupta, 1965). Carboxyl groups were measured by ion-exchange with $Ca(OAc)_2$ solution and potentiometric titration of liberated HOAc with NaOH (Schnitzer and Gupta, 1965). Total hydroxyls were analyzed by acetylation with acetic anhydride in pyridine followed by detection of liberated HOAc by titration with NaOH using phenolphthalein as an indicator (Schnitzer & Khan, 1972). Phenolic hydroxyls were calculated to be equal to the difference between total acidity and carboxyl groups, whereas alcoholic hydroxyls were calculated to be equal to the difference between total hydroxyls and phenolic hydroxyls (Schnitzer & Khan, 1972). Carbonyls were determined by oximation with excess hydroxylamine in methanol-2-propanol and potentiometric titration of unreacted hydroxylamine with standard $HClO_4$ solution (Fritz, Yamamura, & Bradford, 1959).

RESULTS AND DISCUSSION

Chemical and Physical Parameters

Bulk density values ranged from a high of 0.726 g/cc for the surface of Torry to a low of 0.084 g/cc for the 36-to 81-cm horizon of Montverde (Table 2). With increasing profile depth within series, bulk density values decreased, and generally increased with Monteverde < Pahokee < Okeelanta < Torry. Padbury (1970)[3] reported bulk density values of moss-dominated peats ranged from 0.04 to 0.14 g/cc for fibric layers and from 0.14 to 0.26 g/cc for mesic (hemic) layers while sedge-dominated peats ranged from 0.05 to 0.12 g/cc for fibric layers and from 0.12 to 0.30 g/cc for mesic layers. Both peats had bulk densities which ranged from 0.25 to 0.47 g/cc for humic (sapric) layers. Sphagnum moss-dominated peats have bulk density values ranging from 0.05 to 0.1 g/cc for fibric layers and 0.3 to 0.5 g/cc for humic layers (Farnham & Finney, 1965).

The largest ash contents measured at 500C were 67.3% and 70.9% for the two upper horizons of Torry. This ash content is sufficient to qualify Torry for clastic mineralogy. The lowest percent ash measured at 500C was 6.3% for the 36- to 81-cm horizon of Montverde. Ash contents of moss-dominated peats have been reported (Padbury, 1970)[3] to range from 3 to 15% for fibric layer and 15 to 26% for mesic layers whereas sedge-dominated peats ranged from 5 to 12% for fibric layers and 12 to 23% for mesic layers. Both peats had reported ash contents which ranged from 24 to > 60% for humic layers. Frazier and Lee (1971) show ash contents for Wisconsin Histosols of 2.1 to 6.4% for a Fibrist profile, 7.4 to 41.0% for a Hemist profile and 10.8 to 27.5% for a Saprist profile after ignition at 600C. Generally percent ash measured at 500C decreased with increasing profile depth within series, and increased with Montverde < Pahokee < Okeelanta < Torry. Except for the Torry profile, greater percent ash at the soil surface may be due to atmospheric dust but is most likely due to burning, compaction and surface biochemical oxidation. The high ash content of the Torry surface horizons resulted from deposition of overflow sediments from Lake Okeechobee. Similar trends were observed for ash contents determined at 700C with values < 3% lower from those measured at 500C. Differences in ash content determined at 500 and 700C were reported < 2% for fibric layers whereas humic layers showed reductions > 5% (Padbury, 1970)[3].

Ions extractable with neutral 1*N* NH_4OAc were primarily Ca with appreciable Mg and some Na and K. The presence of free carbonates could account for the high extractable Ca and some of the Mg. Sodium concentration increased with depth, probably an expression of its high mobility combined with the presence of a high water table. For crop production these soils are commonly fertilized with 0-10-45 at annual rates of approximately

[3]G. A. Padbury. 1970. Properties of organic soils in relation to classification. Ph.D. Thesis, University of Saskatchewan, Saskatoon, Canada.

Table 2. Selected chemical and physical parameters of Florida Histosols examined.

Series	Depth	Bulk density	Ash		pH		Extractable ions					Saturated conductivity	Water suction (bars)		
			500C	700C	H_2O	1N KC1	Ca	Mg	Na	K	Acidity		0.10	0.33	15
	cm	g/cc	%				meq/100g					cm/hr	% water by weight		
Montverde	0-13	0.258	12.1	10.2	5.8	5.4	103.7	10.8	0.9	3.0	46.7	66.6	251.6	216.9	150.2
	13-36	0.217	10.5	8.5	6.3	5.7	105.2	11.7	1.0	0.5	42.3	87.8	236.8	197.9	137.9
	36-81	0.084	6.3	4.9	6.1	5.6	70.9	8.2	2.7	1.4	43.8	78.4	899.5	648.1	187.6
Pahokee	0-18	0.326	15.2	12.2	6.2	5.6	126.8	12.4	0.6	1.2	39.4	49.1	179.9	151.5	123.1
	18-86	0.176	12.8	11.5	6.2	5.7	141.8	22.1	5.1	0.7	38.0	193.8	459.7	419.0	131.5
Okeelanta	0-21	0.412	45.6	43.9	5.8	5.2	65.5	4.3	0.5	0.9	29.0	67.3	144.8	124.9	97.8
	21-66	0.155	15.3	12.2	5.8	5.2	111.7	10.8	3.5	1.4	60.6	54.8	525.0	407.5	144.7
Torry	0-21	0.726	67.3	65.4	7.0	6.3	76.4	10.4	0.5	3.2	13.6	37.2	77.9	71.7	66.6
	21-53	0.522	70.9	69.7	6.9	6.3	57.5	10.4	0.7	1.9	13.6	16.3	126.7	123.8	121.1
	53-109	0.161	13.7	11.1	6.8	6.2	151.8	29.5	2.7	0.7	35.0	30.9	557.0	459.8	172.6
	109-137	0.131	15.3	14.2	6.3	5.9	128.6	25.9	3.3	0.9	55.9	149.3	532.7	418.2	149.3

560 kg/ha which could account for the higher K levels for surface horizons. Extractable acidity was lowest in the upper two horizons of Torry, which contained most clay, and was appreciably higher in the remaining horizons. Soil reaction in H_2O ranged from pH 5.8 for Okeelanta to pH 7.0 for the surface of Torry. Addition of 1*N* KCl lowered soil reaction in all samples 0.4 to 0.7 pH units.

Saturated conductivity values (Table 2) ranged from a low of 16.3 cm/hour for the 21- to 53-cm horizon of Torry to a high of 193.8 cm/hour for the 18- to 85-cm horizon of Pahokee. Clayton, Neller, and Allison (1942) concluded that water movement in Histosols of the Florida Everglades was much greater in a vertical than horizontal direction. High conductivity values with wide fluctuations are not unusual, since these soils contain partially decomposed sawgrass roots oriented approximately in a vertical position.

Amount of H_2O retained at a tension of 0.10 bar ranged from a high of 899.5% on a weight basis for the 36-to 81-cm horizon of Montverde to a low of 77.9% for the surface of Torry. The largest decrease in water content with a tension increase from 0.33 to 15 bars occurred for the 36-to 81-cm horizon of Montverde which decreased 711.9% water while the smallest decrease occurred for the 21- to 53-cm horizon of Torry which decreased only 5.5% water. Pore size distribution may be more significant in water retention than total porosity. Generally water retention increased with increasing profile depth within series, and decreased with Montverde > Pahokee > Okeelanta > Torry.

Organic Fractions

Generally the humin fraction constituted the greatest organic fraction percentage which ranged from a low of 35.1% for the 21-to 53-cm horizon of Torry to a high of 86.2% for the 36-to 81-cm horizon of Montverde (Table 3). This fraction also increased with depth, and observations of only the surface horizons show Montverde to have the largest humin fraction percentage, Pahokee and Okeelanta to be intermediate and Torry to be lowest. Natural differences in stages of decomposition occur between these soil series in this same order, and within a given soil profile, the surface is usually more decomposed than deeper horizons due to better oxidation conditions.

The humic acid fraction ranged from a high of 54.6% for the 21-to 53-cm horizon of Torry to a low of 7.2% for the surface of Pahokee, while the fulvic acid fraction ranged from 35.8% for the surface of Pahokee to 2.9% for the 35-to 81-cm horizon of Montverde (Table 3). Variations in humic acid or fulvic acid percentages as a function of profile depth or soil series are not apparent although it should be observed that the 36-to 81-cm horizon of Montverde and the 109-to 137-cm horizon of Torry, which are the least decomposed horizons in this study, are lowest in fulvic acid percentages. This phenomenon was also observed from the humic acid/fulvic acid ratios which were high for these two horizons. However, the highest humic acid/fulvic acid ratios occurred in the upper two horizons of Torry. These two horizons

Table 3. Organic fraction distribution and ash content of Florida Histosols examined.

Series	Depth	Organic fraction			Humic acid / Fulvic acid	Ash content: Initial			Ash content: After HC1-HF treatment		
		Humin	Humic acid	Fulvic acid		Humin	Humic acid	Fulvic acid	Humin	Humic acid	Fulvic acid
	cm	%			ratio	%			%		
Montverde	0-13	75.4	13.3	11.3	1.2	6.4	12.6	10.9	3.2	0.8	2.4
	13-36	67.9	15.3	16.9	0.9	7.2	6.0	12.9	2.8	0.9	3.7
	36-81	86.2	11.0	2.9	3.8	2.2	1.2	16.6	2.1	0.6	2.6
Pahokee	0-18	57.0	7.2	35.8	0.2	7.4	6.5	8.2	1.6	2.6	0.9
	18-86	66.8	12.2	21.1	0.6	7.0	4.0	10.4	0.9	0.9	4.5
Okeelanta	0-21	58.7	22.4	18.9	1.2	7.3	4.8	12.5	2.4	0.6	4.4
	21-66	70.1	12.5	17.4	0.7	5.6	4.8	6.9	1.7	0.4	2.2
Torry	0-21	55.4	37.7	6.9	5.4	61.3	36.3	35.4	4.4	0.3	2.2
	21-53	35.1	54.6	10.3	5.3	68.7	65.5	75.6	3.2	0.9	1.5
	53-109	65.3	19.6	15.1	1.3	6.2	7.1	14.7	0.5	0.7	3.3
	109-137	78.4	16.1	5.5	3.0	6.3	8.2	22.5	2.8	0.5	2.0

contain by far the most clay and humic acid. Schnitzer (1967) associated increased humification with increases in fulvic acid but decreases in humic acid concentrations.

Initial ash content of the organic fractions ranged from 1.2% for the humic acid fraction of the 36-to 81-cm horizon of Montverde to 75.6% for the fulvic acid fraction of the 21- to 53-cm horizon of Torry. The largest initial ash contents were observed for the two surface horizons of Torry and except for these horizons was larger in the fulvic acid fraction than either the humin or humic acid fractions. After HCl–HF treatment all ash contents were reduced to $< 5.0\%$.

Elemental Analysis

Examination of the total elemental analysis data revealed striking similarities for all horizons and organic fractions (Table 4). Differences with depth or between series were not apparent. It should be observed, however, that all maximum and minimum values, except for minimum H content, occurred with the first two horizons of Torry. The C content of the humin and humic acid fractions ranged from 56.0 to 61.6% while the fulvic acid fraction ranged from 46.9 to 60.8%. Schnitzer and Khan (1972) reported a 50 to 60% C range for the humin and humic acid fractions and 40 to 50% C range for the fulvic acid fraction. Except for Okeelanta and the lower horizons of Torry, fulvic acid fractions contained a lower C content than the humin or humic acid fractions. The H content of all samples ranged from 3.8 to 7.2%, which is similar to the 3 to 5% H range reported by Schnitzer and Khan (1972) for all organic matter fractions and the characteristic 5% H observed for humic acids (Felbeck, 1971). The N content ranged from 2.3 to 3.9% for the humin and humic acid fractions and from 1.2 to 2.6% for the fulvic acid fractions. These values were similar to the 2 to 4% N range for humin and humic acid fractions and 1 to 3% N range for fulvic acid fractions reported by Schnitzer and Khan (1972).

Variations in C and N analysis of the organic matter fractions were apparent in the C/N ratio which was smallest for the humic acid fractions and largest for the fulvic acid fractions. The fulvic acid fractions contained the largest S content, however, the range for all organic matter fractions was only 0.1 to 1.9%. The O content ranged from 30.5 to 36.4% for the humin and humic acid fractions and 31.1 to 44.3% for the fulvic acid fractions. Schnitzer and Khan (1972) reported a 0 to 2% range of S for all organic matter fractions with an O range from 30 to 35% for humin and humic acid fractions and 44 to 50% for fulvic acid fractions.

Functional Group Analysis

Total acidic groups were generally greatest in the fulvic acid fraction, ranging from 5.1 to 9.8 meq/g and lowest in the humin fraction, ranging from 4.1 to

Table 4. Elemental analysis of humin, humic acid, and fulvic acid fractions from Florida Histosols examined.

Series	Depth	Carbon content			Hydrogen content			Nitrogen content			Sulfur content			Oxygen content			Carbon/nitrogen		
		Humin	Humic acid	Fulvic acid	Humin	Humic acid	Fulvic acid	Humin	Humic acid	Fulvic acid	Humin	Humic acid	Fulvic acid	Humin	Humic acid	Fulvic acid	Humin	Humic acid	Fulvic acid
	cm	%															ratio		
Montverde	0-13	56.9	56.8	53.3	4.4	5.0	4.2	3.1	3.4	2.3	0.2	0.3	1.9	35.4	34.5	38.3	18.6	16.9	22.8
	13-36	56.2	56.2	49.6	4.2	4.6	6.4	3.0	3.4	2.4	0.2	0.3	1.4	36.4	35.5	40.2	18.6	16.7	20.6
	36-81	56.8	58.8	51.2	6.8	4.4	6.4	3.1	3.8	2.6	0.3	0.1	1.1	33.0	32.9	38.7	18.4	15.3	19.4
Pahokee	0-18	56.1	56.3	54.8	4.8	4.0	4.6	3.1	3.6	1.5	0.1	0.1	0.2	35.9	36.0	38.9	18.0	15.8	36.3
	18-86	58.5	56.0	55.3	4.2	5.2	4.4	2.3	2.6	1.3	0.2	0.3	0.3	34.8	35.9	38.9	25.4	21.3	41.9
Okeelanta	0-21	56.4	56.6	56.6	4.4	3.8	4.6	2.7	3.1	2.2	0.3	0.3	0.6	36.2	36.2	36.0	20.6	18.3	26.2
	21-66	57.5	57.8	59.6	4.8	4.2	4.4	2.6	3.3	2.5	0.3	0.4	0.3	34.8	34.3	33.2	22.0	17.6	23.6
Torry	0-21	60.7	56.6	46.9	4.4	4.4	7.2	3.4	3.9	1.2	0.5	0.6	0.4	31.0	34.5	44.3	17.8	14.6	40.8
	21-53	61.6	58.0	52.2	4.2	4.0	6.8	3.4	3.8	1.2	0.3	0.3	0.4	30.5	33.9	39.4	18.2	15.2	45.4
	53-109	58.2	56.6	60.8	4.6	5.0	5.4	2.6	2.8	2.1	0.5	0.4	0.6	34.1	35.2	31.1	22.7	19.9	29.1
	109-137	56.6	57.4	59.4	5.2	5.2	4.2	2.7	3.0	1.6	1.1	0.7	1.3	34.4	33.7	33.5	21.3	19.1	37.4

5.9 meq/g (Table 5). Total acidic groups in the humic acid fraction varied from 4.4 to 8.4 meq/g. Carboxyl groups generally increased in the order of humin < humic acid < fulvic acid. Ranges of carboxyl groups were 1.1 to 3.3 meq/g for the humin fraction, 1.6 to 4.2 meq/g for the humic acid fraction and 2.6 to 4.5 meq/g for the fulvic acid fraction. It was observed that a high carboxyl content especially for the humin and humic acid fractions occurred for the first two horizons of Torry, which contained appreciable clay. Phenolic hydroxyl groups varied from 1.4 to 5.3 meq/g for all organic fractions and were noticeably smaller for the humin and humic acid fractions of the same two horizons of Torry which were high in carboxyl groups. The humin fraction was generally higher in alcoholic hydroxyl groups ranging from 2.0 to 4.8 meq/g than the humic acid or fulvic acid fractions which ranged from 0.3 to 3.3 meq/g and 0.1 to 1.9 meq/g, respectively. Total carbonyl groups were largest for the fulvic acid fraction ranging from 3.7 to 5.0 meq/g, intermediate for the humin fraction ranging from 1.7 to 3.7 meq/g and smallest for the humic acid fraction ranging from 0.5 to 1.9 meq/g. Schnitzer and Desjardins (1966) indicated increased humification for unseparated organic soils was associated with increases in carboxyl, and to a lesser extent carbonyl groups and decreases in alcoholic hydroxyl groups with practically no changes in phenolic hydroxyl groups.

CONCLUDING REMARKS

Although morphologically the Torry surface is highly decomposed, as supported by low humin content and a high humic acid/fulvic acid ratio occurring in the upper two horizons of this series, it has the lowest subsidence rate of all Histosols in the Everglades (Clayton, Neller, & Allison, 1942). Clayton and Neller (1943) reported Torry soils subsiding not more than two-thirds as fast as Pahokee with water tables maintained at the same depth. It appears that decomposition proceeds by biological oxidation with the evolution of CO_2, and the humic acid concentration decreasing with the fulvic acid concentration increasing as a result of humification. Interactions between clay and the humic acid fraction may inhibit humification and hence decrease subsidence. Therefore clay additions may be a feasible mechanism for subsidence abatement.

Schnitzer (1967) indicated that organic matter fractionation is a complex process and high ash contents in the extracts can flocculate the low molecular-weight fulvic acid fraction into the higher molecular-weight humic fraction making reliable estimates of these fractions impossible. The initial ash contents of the humic fractions before HCl–HF treatment for the first two horizons of Torry ranged from 36 to 65% whereas all other horizons were < 13%. If the decreased subsidence rate of Torry is not due to a decrease in humification, as indicated by a low fulvic acid content and a high humic acid content, then it may be related to fulvic acid coagulation by clay which may also decrease subsidence. This suggests that subsidence in

Table 5. Oxygen-containing functional groups of humin, humic acid and fulvic acid fractions from Florida Histosols examined.

Series	Depth	Total acidity			Carboxyls			Phenolic hydroxyls			Alcoholic hydroxyls			Carbonyls		
		Humin	Humic acid	Fulvic acid	Humin	Humic acid	Fulvic acid	Humin	Humic acid	Fulvic acid	Humin	Humic acid	Fulvic acid	Humin	Humic acid	Fulvic acid
	cm	meq/g														
Montverde	0-13	4.3	6.1	9.0	1.6	2.9	4.0	2.7	3.3	5.0	3.2	1.5	1.6	2.0	1.6	4.7
	13-36	4.8	7.0	9.2	1.6	3.4	4.4	3.2	3.7	4.8	2.5	2.0	1.3	2.4	1.3	4.0
	36-81	4.6	4.4	8.6	1.1	1.6	3.8	3.4	2.8	4.8	4.5	2.8	1.9	1.7	1.6	4.1
Pahokee	0-18	5.1	8.4	9.8	1.9	3.2	4.5	3.3	5.3	5.3	4.1	0.5	0.6	2.7	0.9	4.8
	18-86	5.4	8.0	9.1	2.1	3.0	4.1	3.2	5.0	4.9	3.3	1.0	0.6	2.9	0.5	4.2
Okeelanta	0-21	4.1	7.0	8.9	1.7	3.0	4.0	3.4	4.0	4.9	4.6	0.4	0.1	2.4	0.9	3.7
	21-66	5.4	7.7	8.6	1.9	2.7	4.2	3.5	5.0	4.4	3.7	0.3	0.1	1.9	0.8	4.0
Torry	0-21	5.1	7.4	8.1	3.2	4.1	4.4	1.8	3.4	3.7	2.0	1.6	0.1	2.8	1.6	4.4
	21-53	5.3	7.6	9.1	3.3	4.2	4.0	2.0	3.4	5.0	4.8	3.3	0.1	2.9	1.7	5.0
	53-109	5.9	7.6	9.0	1.9	2.5	4.0	4.0	5.2	5.0	3.0	0.4	0.3	2.8	1.9	4.4
	109-137	5.7	8.4	5.1	1.8	3.2	2.5	4.0	5.1	2.6	3.6	0.7	1.9	3.7	1.7	4.2

Histosols low in clay may not be due totally to CO_2 loss but a significant portion may be fulvic acid solubilization with subsequent removal by water. Volk (1972) could account for 73% of the observed subsidence for Montverde as CO_2 evolution, and only 58% for Terra Ceia. It is also possible that the low subsidence rate of Torry is due only to less oxidizable organic matter being available per unit volume as a result of the physical presence of clay.

Similarities in elemental and oxygen-containing functional group analysis between all series and horizons for the humin, humic acid and fulvic acid fractions were surprising. However, it should be realized that these Histosols had very common genetic development over a relatively short period. Initial peat formation began about 4,400 to 5,000 years B.P. (before present) and continued until the early 1900's when extensive subsidence began as a result of the establishment of a drainage program by the Everglades Drainage District (McDowell, Stephens, & Stewart, 1969). Hence elemental and functional group differences should not be as great as those between genetically different series or horizons.

LITERATURE CITED

Clayton, B. S., and J. R. Neller. 1943. Water control investigations. Fla. Agr. Exp. Sta. Annual Report 1943:128–129.

Clayton, B. S., J. R. Neller, and R. V. Allison. 1942. Water control in the peat and muck soils of the Florida Everglades. Fla. Agr. Exp. Sta. Bull. 378. 74 p.

Cooke, C. W. 1945. Geology of Florida. Fla. Geol. Surv. Bull. 29. 339 p.

Davis, J. H., Jr. 1946. The peat deposits of Florida their occurrence, development, and uses. Fla. Geol. Surv. Bull. 30, 247 p.

Farnham, R. S., and H. R. Finney. 1965. Classification and properties of organic soils. Advan. Agron. 17:115–162.

Felbeck, G. T., Jr. 1971. Structural hypotheses of soil humic acids. Soil Sci. 111:42–48.

Frazier, B. E., and G. B. Lee. 1971. Characteristics and classification of three Wisconsin Histosols. Soil Sci. Soc. Amer. Proc. 35:776–780.

Fritz, J. S., S. S. Yamamura, and E. C. Bradford. 1959. Determination of carbonyl compounds. Anal. Chem. 31:260–263.

Jones, L. A. 1942. Soils, geology, and water control in the Everglades region. Fla. Agr. Exp. Sta. Bull. 442, 168 p.

Knipling, E. B., V. N. Schroder, and W. G. Duncan. 1970. CO_2 evolution from Florida organic soils. Soil Crop Sci. Soc. Fla. Proc. 30:320–326.

McDowell, L. L., J. C. Stephens, and E. H. Stewart. 1969. Radiocarbon chronology of the Florida Everglades Peat. Soil Sci. Soc. Amer. Proc. 33:743–745.

Neller, J. R. 1944. Oxidation loss of low moor peat in fields with different water tables. Soil Sci. 58:195–204.

Schnitzer, M. 1967. Humic-fulvic acid relationships to organic soils and humification of the organic matter in those soils. Can. J. Soil Sci. 47:245–250.

Schnitzer, M., and J. G. Desjardins. 1966. Oxygen-containing functional groups in organic soils and their relation to the degree of humification as determined by solubility in sodium pyrophosphate solution. Can. J. Soil Sci. 46:237–243.

Schnitzer, M., and U. C. Gupta. 1965. Determination of acidity in soil organic matter. Soil Sci. Soc. Amer. Proc. 29:274–277.

Schnitzer, M., and S. U. Khan. 1972. Humic substances in the environment. Marcel Dekker, Inc., New York. 327 p.

Soil Survey Staff. 1974. Soil taxonomy: A basic system of soil classification for making and interpreting soil surveys. Soil Conservation Service, USDA Handbook No. 436 U. S. Govt. Printing Office, Washington, D. C. (In press).

Stephens, J. C. 1956. Subsidence of organic soils in the Florida Everglades. Soil Sci. Soc. Amer. Proc. 20:77–80.

Stephens, J. C. 1969. Peat and muck drainage problems. J. Irrig. Drainage Div., Amer. Soc. Chem. Eng. 95:285–305.

Stephens, J. C., and L. Johnson. 1951. Subsidence of organic soils in the upper Everglades region of Florida. Soil Sci. Soc. Fla. Proc. 11:191–237.

Tiedemann, A. R., and T. D. Anderson. 1971. Rapid analysis of total sulfur in soils and plant material. Plant Soil 35:197–200.

Volk, B. G. 1972. Everglades Histosol subsidence 1. CO_2 evolution as affected by soil type, temperature, and moisture. Soil Crop Sci. Soc. Fla. Proc. 32:132–135.

Some Engineering Aspects of Peat Soils[1]

7

I.C. MACFARLANE and G.P. WILLIAMS[2]

ABSTRACT

Peatland covers at least 1,295,000 km^2 of Canada, or about 12% of the total land area and presents major engineering problems, such as access, construction, water and thermal problems. This paper reviews the Canadian approach to these problems and includes an outline of an engineering classification system for peatlands. Geotechnical characteristics of peat are discussed, including certain basic physical properties, settlement characteristics and shear strength. The mechanism of deterioration and corrosion of concrete and metal structures subjected to peatland waters is briefly described. Various factors that determine the ground thermal regime are discussed; freezing and thawing characteristics of peatland are outlined and engineering implications of these characteristics are described. There is a need for continued research on thermal aspects of peat, especially in view of the projected construction of pipelines in the north and the concurrent concern for environmental protection.

INTRODUCTION

No accurate figure is available on the extent of peatland in Canada, but it has been conservatively estimated as 1,295,000 km^2 (500,000 square miles), or about 12% of the total land area of Canada (MacFarlane, 1969). Occurring to a greater or lesser extent in every province and territory, peatland sweeps across a great expanse of the mid-Canada region, overlapping the forest-tundra zone, the boreal forest, and even occurs sporadically in various formations of limited size in the southern, more populated regions of Canada. The major proportion of the peatland habitat, however, falls within the subartic region and within both the continuous and discontinuous permafrost zones.

In Canada, when referring to peatland, the term *muskeg* is frequently used, especially in the engineering literature. This is a term unique to Canada and the northern USA and has its origin in the Algonquin Indian linguistic group. In all dialects, however, it means much the same thing: grassy bog, quaking ground, or swamp. In an attempt to be more precise, the term

[1]Contribution from the National Research Council of Canada, Ottawa.

[2]Research Council Officer, Office of Grants and Scholarships and Research Officer, Geotechnical Section, National Research Council of Canada, Ottawa.

organic terrain was introduced in the early days of research into muskeg problems. This is still used to a certain extent, but the European term *peatland* is now becoming more common.

In the absence of a comprehensive inventory, no accurate maps are available for the extent of peatland in Canada. A map has been developed, however, which gives an indication of the relative severity of the engineering problem, depending upon the frequency of occurrence of peatland (MacFarlane, 1969). Figure 1, a simplified version of this map, shows the distribution of peatland in Canada.

The practical engineering problems presented by the extensive areas of muskeg or peatland, particularly in the subarctic, fall into four general categories, most of which can be discussed only in a cursory manner in this paper.

ACCESS PROBLEMS

Off-Road Access–design and development of off-the-road vehicles, problems of vehicle mobility and terrain trafficability, and the interface between vehicle and terrain. In the past two decades, there has been extensive research and development in Canada in this field, motivated by oil exploration and development in northern Canada and the Arctic. Vehicles are now in production that can carry payloads of up to 27 metric tons with a ground pressure as low as 13.7 kN/m^2 (2 lb/in^2).

GENERAL CONSTRUCTION PROBLEMS

Road and Railway Construction–problems of route selection in peatland areas, stability of embankments, and settlement. Design techniques have been developed in Canada for floating roads (including four-lane expressways) over deep peat deposits. Only minor settlements have occurred after the construction phase.

Buildings–involve settlement problems. The town of Prince Rupert in northern British Columbia, for instance, has many private and commercial buildings sited on deep peat deposits.

Transmission Tower Foundations–stability and settlement problems as well as the design of adequate anchors in peat.

Pipelines–problems of excavation and backfilling, and access of equipment.

Airstrips–problems of embankmant stability and settlement.

WATER PROBLEMS

Drainage–problems of location for drainage ditches, cut-bank stability and general hydrological considerations.

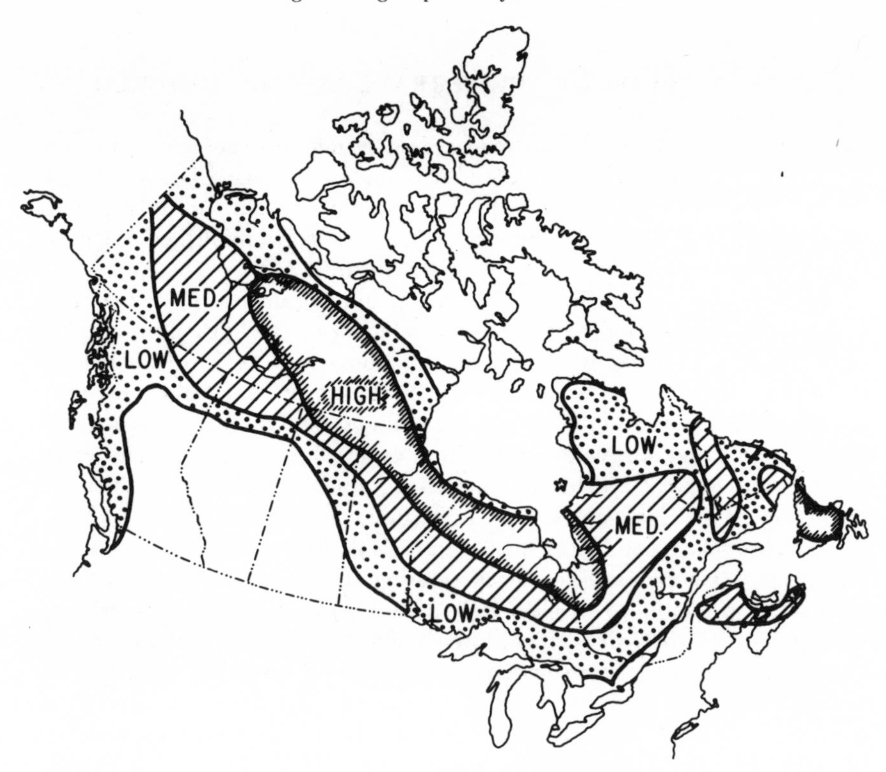

Figure 1. Occurrence of peatland in Canada.

Corrosion—designing protective measures for concrete and metal structures subjected to the aggressive action of peat waters.

THERMAL PROBLEMS

Thermal problems are those involved in continuous and discontinuous permafrost, ground temperatures, and depth of freezing and thawing.

Many of these engineering problems are basically a result of characteristics of both the terrain and the material, that is, their unpredictability, high water table, extremely compressible nature of the material, low shearing resistance, and low bearing capacity.

A great deal of research has been carried out in Canada in the past 25 years by government agencies, universities, and industry in an attempt to understand some of these characteristics of peatlands and peat. From this research has come a classification system for muskeg (peatland) for engineers, as well as a considerable amount of information on the geotechnical properties of peat. The significance of this research to engineering development in peatland areas is shown in this paper.

CLASSIFICATION OF THE TERRAIN AND THE MATERIAL

Motivated by a concern for off-road access by military vehicles in northern Canada, in 1945 the National Research Council encouraged Dr. N. W. Radforth, a palaeobotanist, to develop a classification system for muskeg (peatland) suitable for use by engineers. In 1952 Radforth published *Suggested Classification of Muskeg for Engineers* (Radforth, 1952).

The Radforth classification system for muskeg uses surface vegetation, topographic features and subsurface characteristics. This system utilizes nine pure classes of vegetation designated by the letters "A" to "I" inclusive (Table 1). These letters do not refer to plant species, but rather to certain growth characteristics, such as stature, woodiness, and texture. For instance, Class A represents tall, woody vegetation over 4.8 m (15 feet) high, which, of course, is trees. Class I represents a low, nonwoody plant less than 10.2 cm (4 inches) high, soft and velvety in texture (e.g., moss).

A particular peatland or muskeg area is designated by two or three of the letters in combination, each representing at least 25% of the cover. For example, a common cover class is BEI muskeg, representing dwarfed trees 1.5 to 4.6 m (5 to 15 feet) high, low woody shrubs less than 0.6 m (2 feet) high, and a soft, velvety, nonwoody mat up to 10.2 cm (4 inches) high. The so-called *spruce bog* can be described by this coverage designation. Another common type is FI muskeg, denoting a nonwoody plant type up to 0.6 m (2 feet) high (grass or sedge) and a mossy mat. This type is almost always associated with a very high water table and often with open ponds. Its bearing capacity is very low, often it cannot be safely crossed on foot. The so-called *meadow bog* would fall into the FI classification.

The Radforth classification system also includes the subsurface organic material, or peat. Seventeen peat categories are identified, but they can be divided into three main groups: (i) amorphous-granular peat, (ii) fine fibrous peat (woody or nonwoody), and (iii) coarse fibrous peat. Most engineers tend to describe samples of peat by these simple designations. The relative values of the various peat properties of these three groups are given in Table 2.

Table 1. Properties designating nine pure coverage classes.

Coverage type (class)	Woodiness vs. non-woodiness	Stature (approximate height)	Texture (where required)	Growth habit
A	Woody	4. 8 m or over (15 feet)	--	Tree form
B	Woody	1. 5-4. 8 m (5-15 feet)	--	Young or dwarfed tree or bush
C	Nonwoody	0. 6-1. 5 m (2-5 feet)	--	Tall, grasslike
D	Woody	0. 6-1. 5 m (2-5 feet)	--	Tall shrub or very dwarfed tree
E	Woody	Up to 0. 6 m (2 feet)	--	Low shrub
F	Nonwoody	Up to 0. 6 m (2 feet)	--	Mats, clumps or patches, sometimes touching
G	Nonwoody	Up to 0. 6 m (2 feet)	--	Singly or loose association
H	Nonwoody	Up to 10. 2 cm (4 inches)	Leathery to crisp	Mostly continuous mats
I	Nonwoody	Up to 10. 2 cm (4 inches)	Soft or velvety	Often continuous mats, sometimes in hummocks

Table 2. Relative values of various peat properties for predominant types.

Predominant structural characteristic	Peat property*						
	Water content	Natural permeability	Natural void ratio	Natural unit weight	Shear strength	Tensile strength	Compress-ibility
Amorphous-granular	3	3	2	1	3	3	2
Fine-fibrous (woody and nonwoody)	1	2	1	3	2	2	1
Coarse-fibrous (woody)	2	1	3	2	1	1	3

* On relative scale, 1 is greatest, 3 is least.

The classification system also includes an aerial interpretation system which can be applied for altitudes up to 9.2 km (30,000 feet). Various "air-form patterns" have been defined by Radforth (1955, 1958) from which can be inferred the surface coverage characteristics and, to a certain extent, the subsurface characteristics. This aerial classification system has been used with some success in route selection for access roads in northern Canada, as well as for off-road vehicle routes.

Although the Radforth classification system is not suitable for all purposes, it has been very widely used in Canada by engineers and others and has made a major contribution to the development of northern Canada.

GEOTECHNICAL CHARACTERISTICS OF PEATS

Index Properties

Certain fundamental physical and chemical properties of peat characterize, to some extent, the quality of that peat relative to engineering purposes. As an adjunct to the Radforth descriptive system, the usual practice is to identify a peat by several, fairly easily determined, index properties. These are: water content, specific gravity of soil solids, and either percent ash or organic content. Although these physical properties do not give any clear indication of the strength characteristics of a peat, it is possible to make certain general inferences from them, particularly in relative terms. Table 2 gives the relative values of various peat properties for the three predominant peat types (MacFarlane, 1969). The chemical property of peat most relevant to engineering purposes is, of course, acidity as indicated by the pH measurement of the soil or water.

Water and peat are inseparable by virtue of the very nature of peat formation. Water content is determined by drying the peat at a temperature of 105 to 110C. Although the water content is a very important index characteristic, it does not, by itself, provide a consistent guide to strength unless the degree of humification and mineral contamination are known. In some individual cases, however, a fairly good correlation between water content and shear strength has been found (Figure 2). Some correlation is also indicated between water content and compressibility of peat. In Figure 3, the

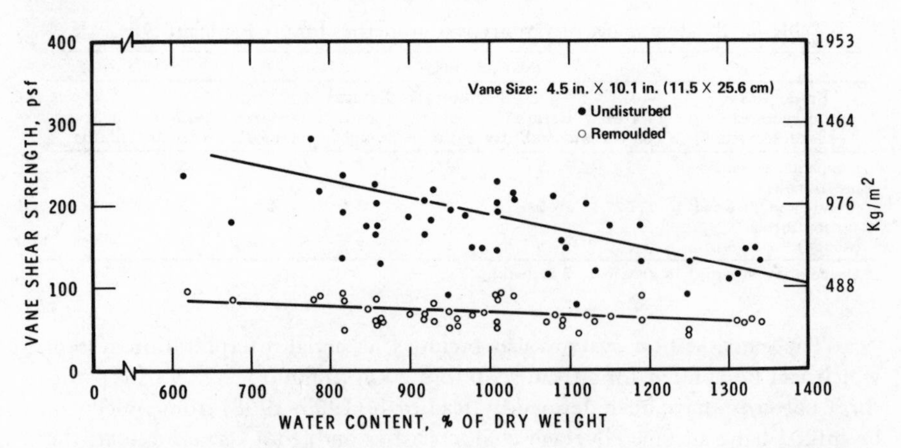

Figure 2. Shear strength vs. water content (Anderson & Hemstock, 1959).

compression index (the larger the compression index, the greater the compressibility of peat) as determined by several investigators is plotted against water content.

The range of specific gravity of peats varies from 1.1 to 2.5 g/cm³ with the average for pure peats being about 1.5 or 1.6 g/cm³. Specific gravity values greater than 2.0 represent peats with a considerable degree of mineral soil contamination. Specific gravity values are used in the calculation of the void ratio (volume of voids/volume of solids), which in turn gives an indication of the compressibility of peat; the higher the void ratio, the greater the potential compressibility.

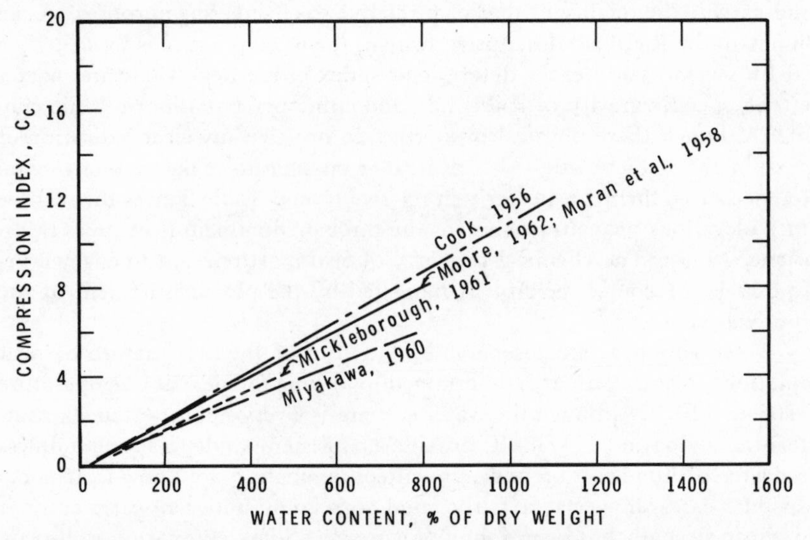

Figure 3. Compression index (C_C) vs. water content (Cook, 1956; Moore, 1962; Moran et al., 1958; Mickleborough, 1961; Miyakawa, 1960).

Organic content is usually determined by burning peat in a muffle furnace at a temperature of 800 to 900°C and determining the percentage of weight loss. Organic content has a considerable effect on the physical and mechanical properties of peat. In general, the greater the organic content, the greater the water content, void ratio, and compressibility of the peat.

Strength and Deformation Properties

SETTLEMENT CHARACTERISTICS

The major involvement of peat in engineering work is in its use as a foundation material. In this role, the high compressibility of the material is the most significant engineering property because large settlements may be caused by relatively small loads. One of the most striking differences in the compression of peat and organic soils as compared with mineral soils is the long-term compression that appears to be an almost continuous process.

Factors that affect the compressibility of peat are the peat type and structure, density, water content, inorganic soil content, gas content, and permeability. The last-noted factor is particularly significant. In the natural state, most peats have a high porosity and are pervious. For this reason the initial compression of peat occurs rapidly. As compression proceeds, permeability is rapidly reduced; even under moderate compressive loading, the permeability change can be several orders of magnitude. An example of this phenomenon was given by laboratory experiments on a specimen of partly humified Irish peat (Hanrahan, 1954) with a natural void ratio of 12 and an initial permeability of 4×10^{-4} cm/sec.

After 2 days under a load of 0.6 kg/cm^2 (8 lb/in^2) the void ratio was reduced to 6.75 and the permeability to 2×10^{-6} cm/sec. After 7 months under the same load, the void ratio was reduced to 4.5 and the permeability to 8×10^{-9} cm/sec, or 1/50,000 of the initial permeability. The change in permeability during compression has, therefore, a profound effect on the rate of consolidation.

The settlement aspect of peat behavior has received a great deal of attention in Canada and extensive literature is now available on the subject as reflected in the Proceedings of the Annual Canadian Muskeg Research Conferences and elsewhere (MacFarlane, 1970). There has been some controversy about the mechanism of consolidation in peat and the relative contributions to the total settlement of the primary consolidation phase and the secondary compression phase. Recent work in the United Kingdom, based to a large extent on Canadian studies, has resulted in the development of rheological models for the consolidation of amorphous granular and fibrous peats, which give theoretical results that seem to be in close agreement with the results of the main investigations carried out elsewhere (Perry & Poskitt, 1972).

Primary consolidation is largely governed by the rate at which water can escape from and through the peat. During this phase, the process is predominantly hydrodynamic. Because of the relatively high permeability of peat in the undisturbed state, primary consolidation develops rapidly; in the laboratory, it generally occurs in a few minutes. Coupled with the normal reduction in hydrostatic pressure associated with the escape of water in the layers under load, reduction in permeability means that consolidation proceeds at an ever-decreasing rate. For some peat types, therefore, the time involved for this first stage of the consolidation process can be considerable and in deep peat may continue for months or even years. Primary consolidation, however, may constitute as little as 50% of the total settlement.

Settlement continues after the hydrodynamic phase at a rate independent of drainage conditions and the thickness of the peat layer. This is known as the secondary compression phase. It is believed to occur simultaneously with the primary consolidation but will continue long after primary consolidation has virtually ceased. The plotted relation between secondary compression and log time approximates a straight line (MacFarlane, 1969). Secondary compression is considered to be of a viscous or plastic nature, its magnitude affected by temperature, the nature of the peat, and the state of stress. It may continue for many years and account for up to one-half of the total settlement.

Embankments and other structures on peat can thus be expected to settle (usually differentially) for a considerable proportion of their existence. Although the greatest amount of settlement takes place in the early life of the structure, difficulties and costly repairs can occur as a result of long-term settlements. To overcome this difficulty, preloading has been successfully used in Canada and elsewhere to eliminate or partially eliminate the long-term compression. A preload of sufficient magnitude and duration is applied to cause the compression of the peat which would normally occur under the proposed design load over the expected life of the structure. The required compression can usually be achieved in a relatively short period of time (months rather than years). Developments of this technique for the construction of major highways over deep peat deposits in British Columbia and Quebec are recorded in detail in the proceedings of Canadian muskeg conferences (MacFarlane, 1970). This technique has made possible the support of not only major highways, but also of oil tanks, buildings, and even small earth dams on peat soils.

SHEAR STRENGTH

The great majority of practical engineering problems involving peat require a knowledge of its shear strength characteristics. The strength of peat is derived from both the fiber strength and the strength of the peat matrix (Helenelund, 1968). Factors that will affect peat strength are; variations in peat structure, water content, and inorganic soil content. To estimate the *in situ* shear strength, a shear vane is used. This is a four-bladed vane which is pushed into the ground to the appropriate depth, rotated, and the torque measured. Knowing the vane dimensions and the torque, the shear strength can be calculated.

Much data on strength, based on *in situ* measurements, has been published in Canada since 1954 (MacFarlane, 1969). It has been found that in certain types of muskeg or peatland the subsurface peat does not have a characteristic shear strength. At peat depths of up to 1.5 or 1.8 m (5 or 6 feet), the *in situ* shear strength will generally be within the range of 488 to 1953 kg/m^2 (100 to 400 lb/ft^2). Anisotropy in strength behavior is of considerable significance in some Canadian peats, however, with the strength in the vertical direction being up to twice as much as that in the horizontal direction (Northwood and Sangrey, 1970). There is evidence that the shear strength results depend upon the size of the vane, but Northwood and Sangrey (1970) report that the optimum vane size is about 10 cm (4 inches) diameter by 20 cm (8 inches) high.

In estimating the maximum height of fill that can be placed on a peat without producing failure, the following approximate formula is used, which is based on experience of many engineers in the field:

$$H_{ult} = 6\tau/\Upsilon$$

where H_{ult} is the ultimate height of the fill, τ is the shear strength of the peat as measured by the vane, and Υ is the unit weight of the fill (Lea & Brawner, 1959).

An accurate estimate of the shear strength is important in designing the height of fill for the preloading technique, to prevent a shear failure in the foundation. Imperfect as it is, the shear vane is the most useful instrument devised thus far for determining the shear strength of peat.

CORROSION OF CONCRETE AND METAL STRUCTURES IN PEAT SOIL AREAS

Owing to the acidic reaction of most peats, concrete and metal structures in such an environment are subject to corrosive attack and precautionary measures may be necessary. Furthermore, the runoff waters from peatland areas have much the same general aggressive properties as the peat and may have a deleterious effect on any structure with which they come in contact. Peatland waters are usually acidic but alkaline waters also occur and may also be corrosive. Running water is potentially more harmful than stagnant water and a constantly changing water level causes more serious attack than when the structure is completely immersed. Frequently the runoff water from peatland areas is brownish in color but the intensity of the color is not a good indication of its potential aggressiveness.

Deterioration of concrete by the action of aggressive waters is a chemical action which is potentially serious if the water is able to percolate through the mass (MacFarlane, 1969). The degree of acidity of the water does not give a simple measure of aggressive action but does bear some relation to it. Deterioration of concrete is also a function of the quality of the concrete since it depends to a large extent on the permeability of the cement paste. The best resistance against attack, therefore, is good quality, dense, concrete with the addition of air-entraining agents for extremely aggressive environments.

Corrosion of metals in a peat soil environment is an electrochemical action. The type and rate of corrosion are functions of the properties of the metal, as well as of the soil and water conditions. Factors such as physical properties of the soil, dissolved salts, pH, total acidity, resistivity, aeration, and presence of anaerobic bacteria influence corrosion. A practical indicator of potential corrosiveness is the soil resistivity, which depends upon the water content of the soil and the dissolved salts in the water. In natural peatland the water content is usually high and the salt content relatively low, with the corrosion potential in the moderate range (Mainland, 1962). Coating and cathodic protection are two of the measures used to counter the effect of an aggressive environment on metals.

GROUND THERMAL REGIME

Factors Determining Ground Thermal Regime

The factors that control the ground thermal regime have some unique characteristics for peatlands; consequently, the thermal regime of peat differs appreciably from that of other soils. These factors include climatic variables (e.g., air temperature and radiation); surface features (e.g., vegetation, microrelief and snow cover); and subsurface factors (e.g., soil moisture and soil thermal properties). The complex interrelationship of these factors with the ground thermal regime can only be discussed briefly in this paper.

CLIMATIC FACTORS

Air temperature over peatland can often be several degrees colder than over adjoining mineral terrain. The drainage of cold air into depressions where peat has formed and the unusual thermal properties of peat are the main factors causing air temperature differences (Williams, 1968). The low air temperatures and associated low surface temperatures over peatland have long been of concern in the cultivation of crops (Bouyoucos & McCool, 1922). It is considered that low air temperatures are also of significance in the occurrence of permafrost (Brown, 1963).

SURFACE FACTORS

Microrelief present in the form of peat mounds, ridges, and plateaux on peatlands influences the occurrence of ice in the ground. The persistence of ice under peat hummocks is frequently observed both in permafrost regions and in regions of seasonal frost. Living vegetation associated with peatlands differs significantly from that over mineral terrain, consisting primarily of a combination of mosses, lichens, and sedges. Vegetation and microrelief also control the depth and distribution of snow, which has a major effect on the depth of frost penetration into the ground in the winter.

WATER CONDITIONS

Soil moisture conditions in peatland control to a large extent the rate of heat transfer from the atmosphere to the soil. Under saturated conditions most of the heat available from solar radiation is expended in evaporation and is not available to warm the soil. During dry weather the surface cover presents high thermal resistance, limiting the amount of heat that can be transferred into the soil.

In studying soil moisture and ground thermal conditions over peatland, surface layers of growing vegetation, mosses, and sedges should be distinguished from the underlying decomposed peat deposits (Romanov, 1961). Water can move freely in the porous vegetative covers by gravity, by capillary action, and by vapor movement. Consequently the moisture content of the surface layer is usually highly variable, responding rapidly to precipitation. The soil water in the underlying decomposed peat is, however, frequently almost immobile with high amounts of water in the soil.

THERMAL PROPERTIES

The somewhat unusual thermal properties of peat soils compared with other soils (Table 3) are major factors in determining the thermal regime of peatland. The thermal conductivity of unfrozen peat, K, at high moisture contents is in the range of the lowest reported values for the mineral soils. Thermal conductivities of dry *Sphagnum* are probably an order of magnitude lower. The volumetric heat capacity, C_v, of peat soil depends almost entirely on moisture content; for saturated peat it is about equal to water. As peat

Table 3. Relative range of values for thermal properties*

	Volumetric heat capacity, cal/cm³℃	Thermal conductivity, millicals/cm-sec-℃	Thermal diffusivity, cm²/sec × 10³	Latent heat of fusion cal/cc
	C_V	K	K/C_V	L
Unfrozen peat	0.6 - 1.0	0.7 - 1.5	1 - 1.5	40 - 80
Frozen peat	0.3 - 0.6	2 - 5	8 - 9	-
Sphagnum (unfrozen)	0.2 - 0.4 (estimated)	0.3 - 0.8	1 - 1.5	15 - 40
Wet sand	0.2 - 0.6	2 - 6	4 - 10	15 - 25
Wet clay	0.3 - 0.4	2 - 5	6 - 16	20 - 30

* Obtained from limited information available in the literature (williams, 1968).

soils have relatively low thermal conductivities and high heat capacities, their thermal diffusivities, K/C_v, are low compared with mineral soils. The latent heat of fusion of peat, L, usually much higher than that of mineral soils, is an important factor in the freezing and thawing of peat soils.

Freezing and Thawing of Peatland

CATEGORIES OF FROZEN PEATLAND

Three categories of frozen peatland have been defined (MacFarlane, 1969). The first, "seasonally frozen," is found south of the permafrost boundary in Canada and, as the name implies, the surface frozen layer is entirely seasonal. The second category is peatland in the discontinuous permafrost zone. This presents the most serious problems to road and other construction because of the erratic distribution of the permafrost. The third category comprises peatlands in the continuous permafrost zone. Peat deposits in this category are generally shallow, their thickness seldom exceeding 0.5 to 1 m. The active layer (the surface layer that freezes and thaws each year) generally varies from about 0.5 to 1.0 m (1.5 to 3.0 feet) in this third category. Its thickness depends on local climatic and terrain conditions but almost invariably extends to the permafrost table, below which the ground is permanently frozen.

DEPTH AND RATE OF FREEZING

The rate and maximum depth of frost penetration in seasonally frozen peatland and in the active layer in permafrost areas is much less than for other soil types under similar climatic conditions. An appreciation of the reason for this can be obtained from the following formula, which is used to estimate the rate of freezing and thawing of soil:

$$x = C\,[(K/L)F]^{1/2}$$

where

C = coefficient that depends on the properties of soil,
F = freezing-index (degree-days),
K = thermal conductivity of soil,
L = latent heat of fusion, and
x = depth of freezing or thawing.

In peat soils K is relatively low and L is relatively high so the depth of freezing is quite low in comparison with that of other soils for a given number of degree-days, F (Brown & Williams, 1972). The maximum depth of seasonal freezing of peat soils seldom exceeds about 0.6 m (2 feet) in the nonpermafrost area of Canada. In contrast, the maximum depths of seasonal freezing of other soils can be 3.1 m (10 feet) or more in the zone of seasonal frost south of the permafrost zone. The depth of freezing of peat bogs with

Sphagnum and appreciable snow cover is often less than 30 cm even during a severe winter.

RATE OF THAWING

The rate of thawing of frozen peat is much slower than mineral soil frozen to the same depth and exposed to similar thawing weather. The reasons are the same as for the shallow depth of freezing (high latent heat and low thermal conductivity) as the same general formula applies for estimating the rate of thaw. It is not uncommon for ice to persist in peat bogs in southern Canada in late June or early July, long after all frost has disappeared from nearby mineral terrain. The slow thawing of cultivated peatland sometimes creates problems because the saturated soil is impossible to work with ordinary farm machinery (Williams, 1966).

Ground Temperatures and Insulating Properties

GROUND TEMPERATURES

Mean annual soil temperatures in peat areas can be from 2 to 3C degrees lower than for other soil types. This difference affects the occurrence of permafrost at sites where the mean annual ground temperature is close to 0C. The depth of penetration of the diurnal and annual temperature waves can be about one-third that of mineral soils (De Vries, 1963), being directly related to the thermal diffusivity of the soil.

PEAT AS A THERMAL INSULATOR

The insulating value of peat has been recognized for many years. It has been used beneath the bearing layer in railway embankments to protect against frost heaving and has also been used in the production of insulating board for the housing industry (MacFarlane, 1969). In permafrost regions the removal of the insulation provided by the organic cover causes an increase in the depth of thaw. It is reported that the maximum depth of thaw in an undisturbed moss-covered area was 0.6 m (2 feet) in contrast to depths of 1.6 to 2.4 m (5 to 8 feet) in areas stripped of their moss cover 3 years previous to the date of measurement (Brown, 1963). Serious structural foundation problems can occur if the underlying permafrost starts to melt because of removal of the peat surface cover.

MODIFICATION OF GROUND TEMPERATURE

In Canada, as elsewhere, there has been considerable interest in the effect of man on the ground temperature of natural terrain such as peatland. In agriculture, modifications to peat soils have often been attempted to improve the thermal regime by changing the thermal properties of soil, utilizing tillage or drainage to make more heat available for warming the soil. Only recently, have these and other methods been considered for such engineering problems as controlling the depth of thaw in permafrost areas. In the USSR various methods of thawing ground for mining purposes have been studied, including removal of peat covers and the drainage of peatlands (Bakakin & Porkhaev, 1959).

Some interesting research work, using mathematical models to predict the effect of surface conditions on the ground thermal regime, has been carried out by the National Research Council of Canada in Ottawa (Gold et al., 1972). These studies included predicting the effect of a layer of peat on the temperatures in the underlying clay and calculating the change in temperatures if the peat layer is replaced by a snow-free gravel fill.

CONCLUSION

This paper has briefly outlined some of the major engineering aspects of peat soils and peatlands, particularly within the Canadian context. Techniques have been developed for successfully building highways and other structures on deep peat deposits. Very large vehicles have been developed capable of carrying heavy loads over peatlands in summer when the ground is unfrozen. Although all of the problems have not been solved there has been sufficient success with some of the practical engineering problems that research into the engineering aspects of peat soils and peatlands has declined somewhat in recent years.

Ecological considerations and the present-day concern for environmental protection means that future research on peatlands in the engineering context will have an entirely different direction. At present, considerable effort is being expended toward the development of a new type of track for large tracked vehicles, which will not break the surface mat and thereby damage the terrain. The probability of oil and gas pipeline construction in northern Canada introduces a new set of engineering problems relative to peatlands, of which thermal aspects loom large. Research in the near future will need to place increased emphasis on the question of the thermal characteristics of peat, especially of perennially frozen peat.

LITERATURE CITED

Anderson, K. O., and R. A. Hemstock. 1959. Relating some engineering properties of muskeg to some problems of field construction. Ass. Comm. on Soil and Snow Mech., Muskeg Res. Conf., Proc. 5th, Nat. Res. Counc. Can.,Ottawa, Tech. Memo 61, p. 16–25.

Bakakin, V. P., and G. V. Porkhaev. 1959. Principles of geocryology, Part II: Engineering geocryology, Chapter V: Principal methods of moisture-thermal amelioration of ing ground over large areas. Acad. of Sciences, USSR, V. A. Obruchev Inst. of Permafrost Studies, Moscow. p. 118–139. (Available as Tech. Trans. 1250, issued by Nat. Res. Counc. Can. Ottawa, Ontario. 1966).

Bouyoucos, G. S., and M. M. McCool. 1922. A study of frost occurrence in muck soils. Soil Sci. 14:383-389.

Brown, R. J. E. 1963. Influence of vegetation on permafrost. Permafrost Int. Conf. Proc., Nat. Res. Counc. Can., Ottawa. p. 20-25. (NRC 9274).

Brown, R. J. E., and G. P. Williams. 1972. The freezing of peatland. Nat. Res. Counc. Can., Div. Bldg. Res., Ottawa. Tech. Pap. 381. (NRC 12881).

Cook, P. M. 1956. Consolidation characteristics of organic soils. Ass. Comm. on Soil and Snow Mech., Canadian Soil Mech. Conf., Proc. 9th, Nat. Res. Counc. Can., Ottawa, Tech. Memo 41, p. 82-87.

De Vries, D. A. 1963. Thermal properties of soils. p. 210-235. *In* W. R. Van-Wijk (ed.) Physics of plant environment. John Wiley & Sons, New York.

Gold, L. W., L. E. Goodrich, W. A. Slusarchuk, and G. H. Johnston. 1972. Thermal effects in permafrost. Canadian Northern Pipeline Res. Conf., Proc., 2-4 February, Nat. Res. Counc. Can., Ottawa. p. 25-45. (NRC 12498).

Hanrahan, E. T. 1954. An investigation of some physical properties of peat. Geotechnique 4(3):108-123.

Helenelund, K. V. 1968. Compression, tension and beam tests on fibrous peat. Int. Peat Congr., Proc. 3rd (Quebec) p. 136-142.

Lea, N. D., and C. O. Brawner. 1959. Foundations and pavement design for highways on peat. Canadian Good Roads Ass., Conv. Proc. 40th (Ottawa, Canada). p. 406-424.

MacFarlane, I. C. ed. 1969. Muskeg engineering handbook. Univ. of Toronto Press, Toronto. 297 p.

MacFarlane, I. C. 1970. Annotated bibliography of engineering aspects of muskeg and peat (to 30 June 1969). Nat. Res. Counc. Can., Div. Bldg. Res., Bibliogr. No. 39. (NRC 11727).

Mainland, G. 1962. Corrosion in muskeg. Ass. Comm. on Geotech. Res., Muskeg Res. Conf., Proc. 8th, Nat. Res. Counc. Can., Ottawa, Tech. Memo 74, p. 100-106.

Mickleborough, B. W. 1961. Embankment construction in muskeg at Prince Albert. Ass. Comm. on Soil and Snow Mech., Muskeg Res. Conf., Proc. 7th, Nat. Res. Counc. Can., Ottawa, Tech. Memo 71, p. 164-165.

Miyakawa, I. 1960. Some aspects of road construction in peaty or marshy areas in Hokkaido, with particular reference to filling methods. Civil Eng. Res. Inst., Hokkaido Develop. Bur. Sapporo, Japan. 54 p.

Moore, L. H. 1962. A correlation of the engineering characteristics of organic soils in New York (preliminary). N. Y. State Dep. of Public Works, Bur. of Soil Mech. Rep. 13 p.

Moran, Proctor, Mueser, and Rutledge. 1958. Study of deep soil stabilization by vertical sand drains. US Dep. Navy. Bur. of Yards and Docks, Rep. No. y88812, Washington, D. C. 231 p.

Northwood, R. P., and D. A. Sangrey. 1970. The vane test in organic soils. Ass. Comm. on Geotech. Res., Muskeg Res. Conf., Proc. 13th, Nat. Res. Counc. Can., Ottawa, Tech. Memo. 99, p. 27-40.

Perry, P. L., and T. J. Poskitt. 1972. The consolidation of peat. Geotechnique 22(1): 27-52.

Radforth, N. W. 1952. Suggested classification of muskeg for the engineer. Eng. J. 35: (11)1199-1210.

Radforth, N. W. 1955. Organic terrain organization from the air (altitudes less than 1,000 feet): Handbook No. 1, Dep. Nat. Defense, Defense Res. Board, DR No. 95, Ottawa, Ont. 49 p.

Radforth, N. W. 1958. Organic terrain organization from the air (altitudes 1,000 to 5,000 feet): Handbook No. 2, Dep. Nat. Defense, Defense Res. Board, DR No. 124, Ottawa, Ont. 23 p.

Romanov, V. V. 1961. Hydrophysics of bogs. (Trans. from Russian by N. Kaner) Israel Program for Scientific Translations, Jerusalem, 1968. TTT67-51286, Washington, D. C. 299 p.

Williams, G. P. 1966. Soil freezing and thawing at the Muck Exp. Sta., Bradford, Ontario. University of Guelph, Eng. Tech. Publ. No. 14.

Williams, G. P. 1968. The thermal regime of a sphagnum peat bog. Int. Peat Congr., Proc. 3rd (Quebec) p. 195-200.

Mapping and Interpretation of Histosols and Hydraquents for Urban Development[1]

8

D.F. SLUSHER, W.L. COCKERHAM, and S.D. MATTHEWS[2]

ABSTRACT

A soil survey of 20,250 ha was made for the Regional Planning Commission. Jefferson, Orleans, and St. Bernard Parishes. Medisaprists, Fluvaquents, Hydraquents, and Haplaquepts were studied. Criteria for soil classification provided by *Soil Taxonomy* (Soil Survey Staff, 1974) resulted in mapping units suitable for many interpretations for urban development. Most soils had severe or very severe limitations for urban uses but differed considerably in their potential for urban development. To provide planners with additional information the soils were ranked in order of their suitability for most urban uses. Much of the New Orleans area is at sea level or below and is dependent upon dikes and pumps for flood protection. Low elevations, poor drainage, subsidence of organic soils, and low bearing strength of semifluid soil material were found to be the main problems in use of soils for urban development. With suitable interpretations the soil survey can be an extremely useful tool for use by planners faced with decisions on urban development of Histosols and Hydraquents.

INTRODUCTION

Urban development in the New Orleans metropolitan area was begun on the higher lying mineral soils on natural levees of the Mississippi River. As population increased, the supply of such soils, locally best suited for urban uses, was exhausted. In 1846 swamp and marsh soils in Orleans Parish were diked and water pumped out (Harrison & Kollmorgen, 1947). The urban area of New Orleans expanded, and 1903 (Rice & Griswold, 1903) maps show encroachment into areas of organic soils.

Population of the New Orleans metropolitan area exceeds 1 million. To provide land for urban and industrial facilities, dikes and levees have been constructed as protection from flooding. Extensive pumping systems have been installed to remove surface water from low elevations. Use of organic

[1]Contribution from the Soil Survey, Soil Conservation Service, USDA.

[2]State Soil Scientist, Soil Correlator, Soil Conservation Service, USDA, Alexandria, La., and Soil Scientist, Soil Conservation Service, USDA, Baton Rouge, La., respectively.

soils for farm, residential, and commercial development has been common for many years but not without many undesirable and costly consequences (Harrison & Kollmorgen, 1947). Residential and industrial development at elevations below sea level are common and increasing. Consolidation and subsidence of organic material has resulted in damage to structures, reduced capacity of drainage systems, and lowering of surface elevation. Because of the low bearing strength of the organic soils and of the mineral soils with high water content, piling must be used in many places to support foundations.

The Regional Planning Commission, Jefferson, Orleans, and St. Bernard Parishes recognized the need for soils information as a basis for planning. In 1968 the Soil Conservation Service entered into a cooperative agreement with the Regional Planning Commission to make detailed soil surveys of 20,250 ha in areas proposed for development during the 1968 to 1988 period.

Many uncertainties faced the survey party. Few detailed soil surveys have been made in swamps and marshes, and the kinds of soil differences and their implication for urban uses were not well-understood. The common characteristics of wetness, flooding, and low elevation, however, were well known to most. There were several major questions. What kinds of useful information could the survey provide for the planning commission? What mapping techniques could be used to traverse and to examine the swamps and marshes to make soil surveys of sufficient detail for urban interpretations? To what extent would classes of *Soil Taxonomy* (Soil Survey Staff, 1974) prove useful in developing concepts of soil series suitable for the possible interpretations? What significant interpretations could be made that would be useful to planners?

THE SURVEY AREA

The New Orleans area is within the Mississippi River Delta. It is bisected by the Mississippi River and bounded by Lake Pontchartrain on the north, Lake Borgne on the east, and Lake Cataouatche on the southwest. Elevations range from about 5 m above sea level on the highest part of the natural levee to 2 m or more below sea level in interdistributary troughs that have been diked and drained by pumps (Figure 1). The natural levees, which occupy the highest elevations, are in bands about 2 km wide on each side of the Mississippi River or in narrow bands that parallel former distributary channels (Fisk, 1960). Soil textures range from loamy to clayey on the natural levees.

Interdistributary troughs are clayey sediments or accumulation of organic material up to 5 m thick. Elevations are near sea level except places where organic soils have subsided after drainage which results in below sea level elevations. The water table is at the soil surface unless artificially drained.

Vegetation is black willow (*Salix nigra*), baldcypress (*Taxodium distichum*), tupelo (*Nyssa aquatica*), and red maple (*Acer rubrum*) in the

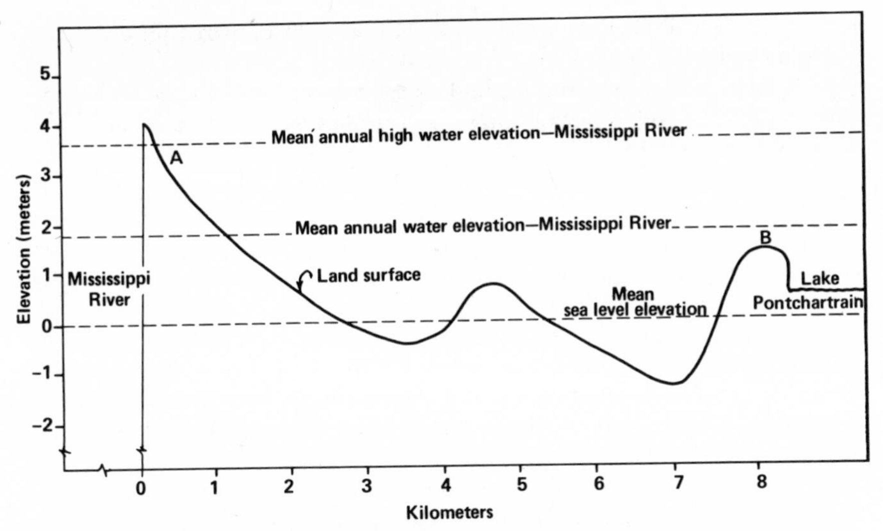

Figure 1. Approximate land surface elevations between the Mississippi River and Lake Pontchartrain along a transect from Jackson Square (A) to University of New Orleans (B).

undeveloped swamp. The freshwater or nonsaline marsh consists of paille fine (*Panicum hemitomon*), waterhyacinth (*Eichhornia crassipes*), pickerelweed (*Pontederia cordata*), alligatorweed (*Alternanthera philoxeroides*), and bulltongue (*Sagittaria lancifolia*). The slightly saline marsh consists of marshhay cordgrass (*Spartina patens*), olney bulrush (*Scirpus olneyi*), saltmarsh bulrush (*Scirpus robustus*), and widgeongrass (*Ruppia maritima*). The marsh is laced with a network of muskrat and nutria trails.

MAPPING AND SAMPLING

Transects across each soil area were planned on the basis of photo interpretation prior to field work. Relocation of transects was done in the field where access was prohibited by locked gates on canals, vegetation in ditches that prevented use of boats, or uncrossable open water bodies on the transect line. Distance between transect lines was 0.5 km to 1 km. Soils were examined at intervals of 60 m along transect lines. Examination was to a depth of 3 m at every other location and 1.5 m at the alternate location. Soil properties were recorded and boundaries located on the basis of field observations, photo interpretation, and transect data.

Soil boundaries were delineated on aerial photographs at a scale of 1:20,000. Photo interpretation was largely unreliable as a tool for separating the Medisaprists from each other and from the Hydraquents due to differences in degree of artificial drainage, burning, and trapping intensity. Separation of the mineral soils on the natural levees was aided by photograph tones.

Copies of the soil maps at a scale of 1:24,000 were prepared for use by the planning commission.

Belgian mud augers and posthole diggers were used to examine organic soils and clayey soils. Bucket augers and probes were used to examine soils in areas dominated by loamy material.

Posthole diggers were the most useful tool for collecting samples for detailed descriptions and laboratory studies from below the water table. Samples were removed by layers using water marks on the posthole digger to get several samples from the same depth. Samples were then laid in sequence on a cloth or plastic sheet for description or subsampling.

Boats were required to cross canals that parallel both sides of most roads in undeveloped areas. Hip boots and waders proved to be too heavy for long foot traverses and were abandoned in favor of light canvas shoes. It was not uncommon for the soil scientist to step in holes concealed by vegetation and sink above waist level.

A soil scientist and an assistant worked together on all transects because of the weight of equipment and in the interest of safety. When possible, two men were dropped off at the start of the transect line and picked up by boat or car at the end of the transect line.

The areas mapped were in scattered tracts that ranged from a few hundred hectares to a few thousand hectares. Smaller tracts took more time per hectare because of time lost in getting to the tracts. Barriers to the mapping process included open bodies of water along foot transects, canals blocked by gates or vegetation that limited access by boat, and buried logs at shallow depth that prevented adequate soil examination. Heavy automobile traffic in a metropolitan area like New Orleans seriously reduces the efficiency of field parties.

SOIL PROPERTIES AND CLASSIFICATION

In the course of field mapping, distinct relationships became evident among soil properties, soil landscapes, and units of classification as set forth in *Soil Taxonomy* (Soil Survey Staff, 1974). Properties used to differentiate soils were mineral content, organic content, consistence as related to water content, and the thickness and sequence of soil layers. Soils that reflected alteration of natural water content by artificial drainage were classified as drained phases of the appropriate taxon. No estimate of the adequacy of flood protection or drainage for urban uses was implied, however, by the phase separation. It was felt that evaluation of adequacy of drainage and flood protection for urban uses was beyond the scope of the soil survey in this kind of terrain.

Most of the mineral soils have very fine textures and montmorillonitic mineralogy. The organic soils are sapric soil material; i.e., highly decomposed. Great groups (Soil Survey Staff, 1974) encountered were Medisaprists, Fluvaquents, Hydraquents, and Haplaquepts (Table 1).

Table 1. Classification, extent, and properties of Great Groups.

Great group	Percent of area	Properties
Medisaprists	43	Dominantly highly decomposed organic soil material (sapric) more than 40-cm thick. Lower horizons may be semifluid or firm clay. High or very high subsidence potential when drained.
Fluvaquents	22	Clayey or loamy mineral soil material. May be semifluid below depths of 50 cm. Lacks diagnostic horizons. Low subsidence potential when drained.
Hydraquents	19	Semifluid clayey soil material to depths of 50 cm or more. May be firm or semifluid below. Lacks diagnostic horizons. Moderate subsidence potential when drained.
Haplaquepts	16	Firm clayey soil material to depths of 50 cm or more. May be firm or semifluid below. Has a cambic horizon. Low subsidence potential when drained.

Interrelationship of n-value, cracking upon drying, and shrink-swell behavior of the clayey mineral soils is illustrated in Figure 2. In the New Orleans area the Sharkey, Fausse, Gentilly, and Barbary soils make up a toposequence of undrained soils (Figure 2). Sharkey soils along with Commerce and Vacherie soils occupy the highest elevations and Barbary soils the lowest. Depth to horizons of low n-value are progressively greater from the higher elevations to the lower.

Important properties of Medisaprists and Hydraquents with Histic epipedons are illustrated in Figure 3. The unnamed Hydraquent has a thin surface of organic soil material over semifluid clay. The Allemands soils have a moderately thick layer, and the Kenner, Lafitte, and Maurepas soils have thick layers of organic soil material. The organic soil material of the Kenner soils is interspersed with thin clay lenses that are absent in the Lafitte and Maurepas soils. Maurepas soils formed from woody vegetation whereas Allemands, Kenner, and Lafitte soils formed from herbaceous vegetation.

DISCUSSION

Historically the mineral soils on the natural levees were the first ones to be developed for urban uses. These areas were soon exhausted, and the press of increased population resulted in expansion into areas of semifluid mineral soils and organic soils. The severity of soil limitation was given little consideration (all soils were built on). Buildings were supported by pilings.

The soil survey points out differences in limitation of soils for urban development. The differences are related to observable or measurable soil properties or groups of properties. Taxonomic classes which are based on sets of properties provide useful and practical soil series.

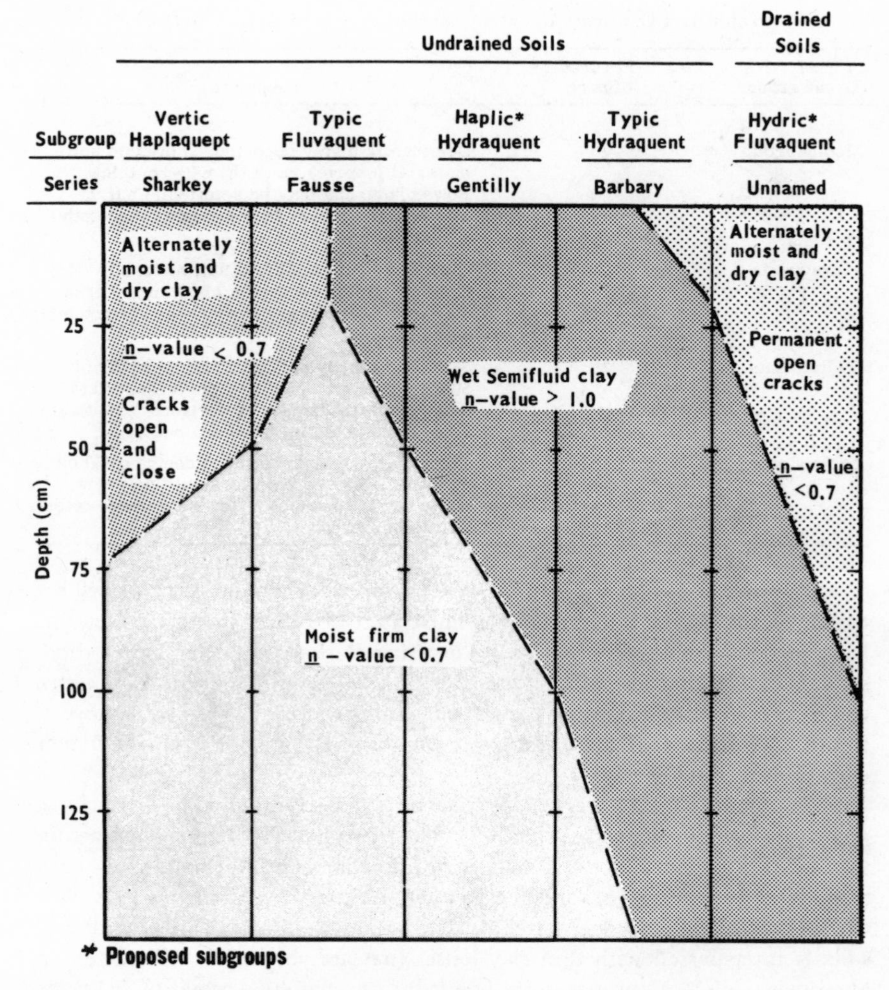

Figure 2. Relationship and properties of mineral soils.

Organic Versus Mineral Soil Material

Whether soil material is mineral or organic is of utmost importance in predicting potential for urban use. Organic soil material, as defined in *Soil Taxonomy* (Soil Survey Staff, 1974) could be identified quite reliably by field estimation and soils placed in appropriate taxa of Histosols. Organic soils have low bulk density, are highly compressible, shrink irreversibly when drained (Figure 4), undergo continual subsidence after drainage due to biochemical oxidation, are subject to fire, have low bearing strength and, in the New Orleans area are below sea level elevations after drainage.

Some soils are moderately consolidated in their natural state; others are semifluid but become consolidated after drainage. Most have severe limita-

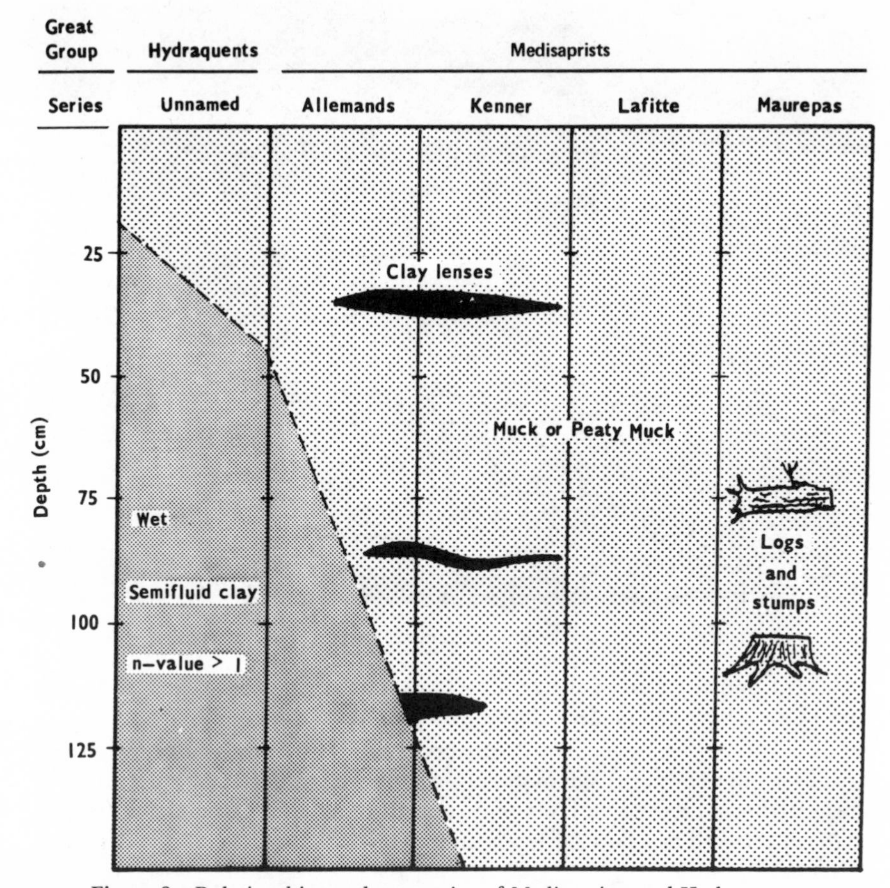

Figure 3. Relationships and properties of Medisaprists and Hydraquents.

tions for building sites in the New Orleans area because of extreme wetness and high plasticity. Mineral soils are favored however because of better engineering properties and they undergo relatively little subsidence after drainage. Mineral soils were classified as Fluvaquents, Hydraquents, or Haplaquepts.

Soil Consistence

The attributes of soil material that are expressed by the degree and kind of cohesion and adhesion or by the resistance to deformation or rupture are termed consistence. Many of the soils were so soft and so wet that their consistence was beyond the scale given for wet consistence in the *Soil Survey Manual* (Soil Survey Staff, 1951), therefore, additional ad hoc classes were defined and used. Consistence for the soft wet soils was described in terms of the ease with which a mass of soil flowed between the fingers when

Figure 4. Large cracks in levee constructed from Kenner muck.

squeezed and the size of the residue remaining in the hand. The following classes for mineral soil material were recognized: (i) soil material that flowed easily between the fingers when squeezed and left no residue or only a small residue in the hand, (ii) soil material that flowed with difficulty between the fingers when squeezed and left only a small residue in the hand and (iii) soil material that did not flow between fingers, or that flowed with difficulty and left a large residue.

Mineral soil material that flowed easily between fingers when squeezed and left only a small residue in the hand was found to have water content in excess of 100%[3]. Semifluid is the term used to express this condition. Mineral soils of this consistence are classified as Hydraquents (Soil Survey Staff, 1974). When drained, clayey Hydraquents shrink, and cracks 5 cm to 15 cm wide form at the surface. Much of the shrinkage is irreversible and cracks remain when the soil is rewet.

The consistence of semifluid mineral soils can be expressed as the *n*-value (Soil Survey Staff, 1974). The *n*-value can be calculated for mineral soil materials that are not thixotropic by the formula

$$n = (A - 0.2R)/(L + 3H),$$

where A = percentage of water in a soil at field condition, calculated on a dry weight basis; R = percent silt plus sand; L = percent clay; and H = per-

[3]Unpublished data, Soil Survey Laboratory, Soil Conservation Service, USDA, Lincoln, Nebraska.

cent organic matter (organic carbon × 1.724). If the soil flows easily between fingers when squeezed, leaving only a small residue in the hand, the *n*-value is estimated to be 1 or more. If the soil does not flow between fingers, or flows with difficulty and leaves a large residue, the *n*-value is less than 0.7. Most soil material deposited under water and never dried below field capacity have *n*-values of 1 or more. The *n*-value is helpful in indicating trafficability in the undrained state and the degree of subsidence that would occur following drainage. With practice, soil scientists become skilled at estimating *n*-values just as proficiency is achieved in estimating particle size and textural classes.

Hydraquents must be drained and allowed to consolidate to make suitable building sites. The barbary and Gentilly series are Hydraquents. Barbary soils become firm to a depth of about 1 m after drainage (Figure 2) but remain semifluid beneath. They are used for building sites but pilings are required to support most structures. A new subgroup (Soil Survey Staff, 1974), Hydric Fluvaquents, is proposed to accommodate Fluvaquents having *n*-values of more than 0.7 in some layers below 50 cm but within 1.25 m. The Gentilly soils are Hydraquents[4] with semifluid layers to a depth of at least 50 cm but are firm below 1 m. They are believed to result from deposition of semifluid soil material over older soils that have consolidated before the later deposition. After drainage the Gentilly soils are firm throughout and are considered more desirable for homesites than Barbary soils. A new subgroup (Soil Survey Staff, 1974), Haplic Hydraquents, is proposed for Hydraquents having *n*-values less than 0.7 below depths of 50 cm but within 1.25 m.

Subsidence

The loss of surface elevation after a soil with organic or semifluid mineral layers is artificially drained (Stephens & Speir, 1969) is termed subsidence. Subsidence of organic soils after drainage is attributed mainly to four factors: (i) shrinkage due to desiccation, (ii) consolidation by loss of the bouyant force of ground water and by loading, or both, (iii) compaction due to tillage, and (iv) biochemical oxidation. Elevation loss due to the first two factors is termed *initial subsidence* and is normally accomplished in about 3 years after lowering the water table. Oven-dried samples of organic layers of a Kenner soil lost 85% of the original volume. Artificial drainage however does not reduce water content to that extent, and volume change under field conditions is less. Initial subsidence or organic soils is estimated to result in a reduction of thickness of the organic materials above the water table by about 50%, and it is accompanied by permanent open cracks that do not close when the soil is rewet. After initial subsidence, shrinkage will continue at a fairly uniform rate due to biochemical oxidation of the organic materials. This is

[4]Gentilly soils were classified as Haplaquepts in this survey but have been reclassified as Hydraquents.

termed *continued subsidence* and progresses until mineral material or the water table is reached. The rate of continued subsidence depends upon temperature (number of days per year above 5C), the mineral content, and depth to water table. The rate increases with depth to the water table. The average rate in the New Orleans area is estimated to be 1 to 5 cm/year.

Continued subsidence is a severe limitation to land use in the New Orleans area, especially if structures are built. Unless piling is used to support buildings, they tilt and foundations crack. Soil around structures built on piling subside at a slow rate. Foundations are exposed and unsupported driveways, porches, and sidewalks crack and gradually drop below original levels (Figure 5). Driveways have subsided to the extent that it is not possible to drive into a carport. Fireplugs stand above the original levels and concrete foundations of utility poles, originally below the surface, are exposed as much as 1 to 2 m above the surface. Subsidence takes place gradually, perhaps only as much as 1 to 5 cm/year, but the total subsidence potential is as much as 4 m for some soils in the New Orleans area.

Subsidence of organic soils can be prevented by maintaining the water level at the surface. It can be reduced by maintaining the water level as high as possible for the land use. Subsidence can also be reduced by covering the surface with mineral soil material to slow diffusion of oxygen and reduce thickness of organic material between mineral soil material and the water table. In land use decisions, a choice must be made of (i) using the land without drainage, (ii) using the land with some drainage but tolerating wet condi-

Figure 5. Settlement of concrete work around house due to subsidence of organic layers of a Medisaprist.

tions and little or no subsidence, (iii) providing adequate drainage and tolerating the subsidence, or (iv) bearing the expense of covering the surface with mineral soil material to reduce subsidence. The thickness of fill required or its effectiveness in reducing subsidence has not been fully evaluated, and further study is needed. Homeowners in the New Orleans area periodically haul fill material to spread around foundations in an attempt to maintain original ground levels.

Mineral soils with the semifluid layers (n-value of 1 or more) have a potential for initial subsidence due to loss of water and consolidation after drainage but have little if any subsidence thereafter.

Fire

A constant hazard in areas of drained organic soils is fire. Fires are difficult to extinguish, leave irregular land surfaces, produce large quantities of smoke, and are a threat to structures. They were observed only in undeveloped areas.

Logs and Stumps

In the Barbary and Maurepas soils, logs and stumps up to 2 m in diameter were consistently found at depths of 0.8 to 1.5 m beneath the surface. They affect the ease of excavation and the uniformity of subsidence when drained. Soil surface conditions or existing vegetation were not reliable indicators of buried logs and stumps because of encroachment of herbaceous vegetation into former swampland. Volume estimates were impossible to make with the field techniques that were used. Frequency of borings were recorded in which large wood fragments prevented further penetration by the auger.

DESCRIPTION OF SOILS AND RATINGS OF LIMITATIONS

The soils were described and interpretations prepared in a special report for the Regional Planning Commission (Soil Survey of Portions of Jefferson, Orleans, and St. Bernard Parishes, Regional Planning Commission, New Orleans, Louisiana, 1970). Soils were described by narrative at the series level, by a range of characteristics for each series, by a mapping unit narrative, and by a typical profile for each mapping unit. Mapping unit descriptions showed the soil as encountered in mapping. Mapping units that were undrained at the time of mapping were also discussed under a separate subheading of, "If protected and drained." This discussion explained the effect of drainage on the soil and benefits and problems likely to result from drainage.

Engineering interpretations were prepared according to the *Guide for Interpreting Engineering Uses of Soils* (Soil Conservation Service, USDA,

Table 2. Engineering and other selected use interpretations.

Soil group and description	Soil name	Homesites and light industry*
A. Mineral soils with consolidated clayey or loamy layers throughout that typically occur at the higher elevations in the area.	Commerce silt loam	Protected MODERATE--moderate wetness
Low subsidence potential, low to very high shrink-swell potential, most foundation designs for one- or two-story buildings do not require piling. Moderate to severe degree of limitation for most urban uses after drainage.	Sharkey clay	Protected SEVERE--very high shrink-swell potential, severe wetness, medium bearing strength
B. Mineral soils with thin organic surface layers and organic soils that are underlain by consolidated clayey layers. Medium to high subsidence potential, very high shrink-swell potential and low bearing strength. Most foundation designs for one- or two-story buildings require piling. Severe degree of limitation for most urban uses after drainage.	Gentilly muck	Protected SEVERE--very high shrink-swell potential, medium subsidence potential, severe wetness, low bearing strength Unprotected VERY SEVERE--flooding
	Allemands peat, firm substratum	Protected SEVERE--high subsidence potential, severe wetness, moderate to severe fire hazard, very high shrink-swell potential, low bearing strength Unprotected VERY SEVERE--flooding
C. Mineral soils with clayey or thin organic surface layers underlain at 76 cm to 1 m (30 to 40 inches) with semifluid clayey or organic layers.	Sharkey clay, miry subsoil variant	Protected SEVERE--very high shrink-swell potential, severe wetness, low bearing strength
Low to medium subsidence potential, none to slight fire hazard, low bearing strength, logs and stumps. Foundation designs for one- or two-story buildings require piling. Severe degree of limitation for most urban uses after drainage.	Barbary soils	Protected SEVERE--very high shrink-swell potential, medium subsidence potential, severe wetness, slight fire hazard, low bearing strength Unprotected VERY SEVERE--flooding

(continued on next page)

D. Organic soils with 38 cm to 1.3 m (15 to 50 inches) of organic material over semifluid clayey layers. High subsidence potential, moderate to severe fire hazard, very high shrink-swell potential, low bearing strength. Foundation designs for one- or two-story buildings require piling. Severe limitations for most urban uses after drainage.	Allemands muck, drained	<u>Protected</u> SEVERE--high subsidence potential, severe wetness, severe fire hazard, very high shrink-swell potential of mineral layers, low bearing strength
	Allemands peat	<u>Protected</u> SEVERE--high subsidence potential, very high shrink-swell potential of mineral layers, severe wetness, moderate to severe fire hazard, low bearing strength <u>Unprotected</u> VERY SEVERE--flooding
E. Organic soils with 1.5 to 3.7 m (5 to 12 feet) of herbaceous organic materials over semifluid clayey layers. Very high subsidence potential, severe fire hazard, low bearing strength. Foundation designs for one- or two-story buildings require piling. Very severe limitations for most urban uses after drainage.	Kenner soils, drained	<u>Protected</u> VERY SEVERE--very high subsidence potential, severe fire hazard, low bearing strength
	Lafitte muck	<u>Protected</u> VERY SEVERE--very high subsidence potential, severe fire hazard, severe wetness, low bearing strength <u>Unprotected</u> VERY SEVERE--flooding
F. Organic soils with 1.5 to 3.7 m (5 to 12 feet) of woody organic materials that have layers of logs, stumps and wood fragments over semifluid mineral layers. Very high subsidence potential, severe fire hazard, low bearing strength. Wood fragments, logs and stumps result in irregular subsidence. Foundation designs for one- or two-story buildings require piling. Severe or very severe limitations for most urban uses after drainage.	Maurepas muck, loggy	<u>Protected</u> VERY SEVERE--very high subsidence potential, low bearing strength, severe wetness <u>Unprotected</u> VERY SEVERE--flooding

* Ratings for <u>Homesites and light industry</u> assume community sewage systems. Protected is with levees and pumpoff drainage if needed. Unprotected is subject to flooding.

1967). The degree of soil limitation for several uses was given in tabular form. Serious deficiencies were encountered in the use of standard limitation ratings because according to nationwide criteria, soils with high potential for urban uses in the New Orleans area were rated the same as soils with fair to low potential. Sharkey clay (Vertic Haplaquepts) having severe limitations due to high shrink-swell behavior is rated the same as Allemands muck, drained (Terric Medisaprists) having 0.4 to 1.25 m of organic material in the upper layers. The shrink-swell behavior is relatively easy to overcome by special design of structures, whereas no feasible solution to the problem of subsidence of organic soils is known. Both soils have severe limitations—one is correctable however, the other is not.

To provide users of the survey with further evaluation of soil suitability for urban uses, the soils were arrayed by groups and within major groups in order of increasing limitations for most urban uses after drainage and protection from flooding. The 19 mapping units were placed in 6 groups for use in preparing interpretive maps. Characteristics considered in grouping the soils were (i) consistence of mineral layers, (ii) contents of mineral and organic matter, (iii) thickness of organic layers, and (iv) presence of buried logs and stumps. For most urban uses soils placed in group A are the most desirable for urban uses even though limitations range from moderate to severe. Soils in each succeeding group have limitations more difficult and more costly to overcome even though the rating for degree of limitation might be the same for soils in each group. As a further refinement, soils within each group are arrayed in order to increasing limitations. To illustrate the format used, an excerpt from the special report is given in Table 2. Further improvement by ratings of development potential using costs of development and limitations after development would make the soil survey information easier for planning officials to use.

SUMMARY AND CONCLUSIONS

Population growth in metropolitan areas has resulted in urban development on soils with very severe limitations for residential and industrial uses because of their proximity to existing facilities. When planning officials are faced with land use decisions between soils, all of which have severe limitation ratings by standard nationwide criteria, the present conventions for presenting soil limitation ratings need to be augmented so as to set forth important interpretive differences between such soils.

Ratings of soils according to costs of development and soil limitations remaining after development offers possibilities for improving the use of soil survey information. Planning officials must realize too that it may not be economically feasible to develop and maintain suitable building sites on thick organic soils with present inadequate technology for control of subsidence.

Soil properties considered most significant in evaluating the limitations of soils for urban development in the survey area were content of mineral and organic matter, thickness of organic layers, consistence as related to water content (*n*-value), and presence of buried logs and stumps. Evaluation of adequacy of artificial drainage systems, flooding hazard, and elevation, although important to land use decisions, were considered beyond the scope of the soil survey even though some soils were classified as drained phases.

The criteria for classes in *Soil Taxonomy* (Soil Survey Staff, 1974) were useful and practical as a basis for developing soil series concepts. The need was recognized for additional subgroups to accommodate Haplaquents and Fluvaquents with high *n*-values below 50 cm and Hydraquents with low *n*-values within 1.25 m. Drainage of Hydraquents results in a change of classification of Fluvaquents or Haplaquepts.

Satisfactory interpretation for an intensive land use such as homesites requires classification of the soils at least to the family level, and in some cases the series must be phased to accommodate significant differences.

Special equipment, such as all-terrain vehicles or helicopters, is needed to traverse marshes and swamps to examine soils at intervals close enough to prepare soil surveys in sufficient detail for intensive land use planning.

LITERATURE CITED

Fisk, H. N. 1960. Recent Mississippi River sedimentation and peat accumulation. Compte Rendu du quatrieme Congres pour l'avancement des e'tudes de Stratigraphie et de Geologie du Carbonifere. Heerlen, Netherlands. p. 187-199.

Harrison, R. W., and W. M. Kollmorgen. 1947. Past and prospective drainage reclamations in the coastal marshalnds of the Mississippi River Delta. J. Land Public Util. Econ. 23(3):297-320.

Rice, T. D., and L. Griswold. 1903. Soil Survey of the New Orleans Area, Louisiana. USDA Bureau of Soils.

Soil Conservation Service, USDA. 1967. Guide for interpreting engineering uses of soils. SCS-USDA, Washington, D. C.

Soil Survey Staff. 1951. Soil survey manual. USDA Handbook No. 18.

Soil Survey Staff. 1974. Soil Taxonomy: A basic system of soil classification for use in making and interpreting soil surveys. Soil Conservation Service, USDA. Agriculture Handbook No. 436. (In press). (Information taken from selected chapters of the unedited text.)

Stephens, J. C., and W. H. Speir. 1969. Subsidence of organic soils in the USA. Association Internationale D'Hydrologic Scientifique. Extract de la Publication No. 89. Colloque de Toyko.

Use of Organic Soils for Wastewater Filtration[1]

9

R.S. FARNHAM[2]

ABSTRACT

Investigations using organic soils for wastewater renovation have been conducted in Minnesota for the past few years. Use of naturally occurring organic soils, as well as organic materials spread over sand have both been effective in reducing phosphate and organic matter of municipal wastewater. These investigations have included design criteria for filtration systems, rates of wastewater application and efficiency of the various combinations of filtering materials. Phosphorus in wastewaters was effectively removed in the orthophosphate form using organic soil materials and wastewater with a high carbon to phosphorus ratio. The immobilization of phosphate has been attributed to the high bacterial population. The biochemical oxygen demand (BOD) load of the wastewater was very effectively reduced (almost 100% reduction) by physical sorption on fibers and colloids of the organic filtering media and through biological decomposition although application rates were high as 20 cm/day. Grass crops growing on the organic soil surfaces were effective in removing nitrates from the wastewater.

INTRODUCTION

The adverse effects of nutrient and organic matter pollutants from wastewater treatment systems on the water quality of many of our lakes and streams has been well documented in the recent literature. Phosphates in particular have been cited as the principal nutrient causing undesirable algal blooms in lakes (Taylor, 1967). Conventional wastewater treatment systems, although quite effective in removing organic matter or biochemical oxygen demand (BOD), have been very ineffective in nitrogen and phosphorus removals.

One approach to solving this problem is treatment of wastewater by application of soil systems (land treatment) either with a cover of trees or crops especially grasses. This study is concerned with the feasibility of utilizing organic soils under specified conditions for the effective removal of pollutants in wastewater. Although little information is available on the use of organic soils for wastewater treatment there are many published reports avail-

---

[1]Contribution from the Soil Science Dept., University of Minnesota and the Minnesota Agricultural Experiment Station as Scientific Journal Series no. 8554. Much of the financial support for this research was from the Environmental Protection Agency and the project was in cooperation with the Iron Range Resources, State of Minnesota.

[2]Professor of Soil Science, University of Minnesota, St. Paul, Minnesota.

able describing the use of mineral soils as treatment systems utilizing either irrigation techniques or rapid soil infiltration basins. Notable among these are the Pennsylvania studies of Parizek et al., 1967; Day, Tucker, and Vanich, 1962; Bouwer, 1968; and a recent book which is a complete review of the subject by Temple University (1972).

The use of organic soils for treating wastewater has been reported in Finland by Surakka and Kamppi, 1971. They applied raw and secondary sewage effluents to ditched peat areas and treatment was effected when the wastewater infiltrated the peat between ditches. The reduction of phosphorus was 82% and nitrogen 90% at one location. Another paper from Finland by Silvo (1972) reported on the use of sphagnum peat for purification of slaughter house wastewaters. An earlier paper by Henry et al. (1954) describes the use of organic soils for treating sewage effluent through crop irrigation using reed canarygrass (*Phalaris arundinacea* L.). Rogstad and Larson, (1969) showed an 88% reduction in orthophosphate of stabilization pond effluent after filtering through organic soils. In a recent report Farnham and Brown (1972) describe a study of organic soils using both lysimeters and leaching columns for the treatment of secondary municipal wastewaters. They reported the removal of over 2,000 kg/ha of phosphorus during a 12-month period using high rates of application. Biochemical oxygen demand was reduced from an average of 140 to 160 mg/liter in the wastewater to less than 2 mg/liter in the filtered effluent. Coliform bacteria were reduced from an average of 1×10^7 bacteria/100 ml to $< 100/100$ ml in the discharge. They also reported nitrogen removal by quackgrass as high as 896 kg/ha from a dry matter yield of 14 to 18 metric tons/ha. These yields using irrigated wastewater are much higher than the average grass yields obtained by farmers in northern areas using normal agronomic practices.

The objectives of this research were to evaluate the effectiveness of the principal organic soil types in (i) removal of phosphorus and organic matter (BOD reduction) of wastewater applied by overhead irrigation, (ii) develop a practical and economical waste treatment system using organic soils, and (iii) utilize several types of grasses to maximize the uptake of both nitrogen and phosphorus contained in the wastewater.

MATERIALS AND METHODS

The methods of study used in these investigations were described in detail by Farnham and Brown (1972). Briefly, they included the use of both outdoor lysimeters and acrylic plastic cylinders (soil columns) in the laboratory for determining the hydraulic properties and nutrient removal efficiencies of the three organic soil materials. The organic soil materials used have the following characteristics:

1) Fibric (Sphagnum Moss type)–pH = 3.5, ash content = 2.5%, bulk density = .06 to .08 g/cc (3.7 to 5.0 lb/cu ft), saturated water content = 95.7% (vol. basis, oven dry).

2) Hemic material–pH = 4.8, ash content = 8.5%, bulk density = 0.17 to 0.20 g/cc (10.6 to 12.5 lb/cu ft), saturated water content = 86.0% (vol. basis, oven dry).
3) Sapric material–pH = 6.0, ash content = 12.0%, bulk density = 0.25 g/cc (15.6 lb/cu ft), saturated water content = 82.5% (vol. basis, oven dry).

The ultimate design of the filter system was largely determined by trial and error. Many variables such as the kind and thickness of organic soil material, the use of sand to dewater the overlying organic material, bulk density of the media, the rates and frequency of applying wastewater, and seasonality of operation were all investigated during the course of the studies.

The optimal thickness of organic soil material was found to be 30 cm since this provides an adequate root zone for grass plants.

Fine sand was used as the media below the organic materials in order to dewater them to a level which would permit sufficient air pore space throughout the entire thickness of the organic soil.

The bulk density of the upper organic layers were artificially created by careful packing so as to equal that occurring in the natural condition in peatlands. Care was exercised to avoid air pockets at the contact of the organic soils with the sand below.

Application rates varied from 5 to 20 cm/day but it was determined later that the most effective rates were those that maximized grass production and hence increased N and P removal.

Some of the lysimeters contained no vegetation and others had a cover of either native sedges and grasses or were seeded to quackgrass (*Agropyron repens*). Cylinders in the laboratory had a cover of either bluegrass (*Poa pratensis*) or quackgrass.

Samples of the discharge effluent from both the cylinders and lysimeters were collected weekly and analyzed routinely for pH, BOD, and phosphate and occasionally for nitrate nitrogen and coliform bacteria content. The analytical procedures used were those recommended by the American Public Health Ass. (1971).

The wastewater used in all studies was obtained from a secondary municipal treatment plant. It had the following characteristics: pH 7.0 to 7.5; phosphate as P varying from 6 to 9 mg/liter and an average BOD of from 110 to 140 mg/liter; nitrate nitrogen varied seasonally and ranged from 10 to 20 mg/liter as N.

Four outdoor lysimeters measuring 3 m (10 feet) in length by 0.5 m (5 feet) in width and 0.5 m (5 feet) in depth were used. A plastic liner was placed inside the wooden lysimeters and outlet pipes were placed in gravel at the bottom for drainage and sample collection.

A small automatic plot irrigator was used to apply wastewater to the lysimeters. This device had a movable boom with spray nozzles pointing downward and valves were used to control the rate of application. A water meter was installed in the line to record the amount of effluent applied during a given time.

RESULTS AND DISCUSSION

The fibric, hemic, and sapric organic materials all proved to be suitable media hydraulically and functioned satisfactorily for nutrient removal even under excessive rates of application of wastewater. The hemic material was the best growth media for the grasses used but none of these organic materials became clogged due to additions of organic matter even after 1 full year's operation. If anything the structure of the surface soil improved with time and the infiltration capacity actually increased.

Table 1 shows the equilibrium water content after saturating hemic organic material over sand and over a screen. The air pore spaces throughout the entire depth of the 30 cm of hemic material over 60 cm of fine sand are relatively high; 29 to 31% of total volume. On the other hand, a 30-cm column of this same organic soil material over a screen has much less air pore volume even at the surface and at the bottom the soil is completely saturated with water and contains zero air pore space. The fine sand, which has finer pores than the organic soil, exerts a tension or suction on the coarser organic soil material above and reduces its water content to a desirable range. This maximizes the aerobic biological processes and accelerates the decomposition of the added organic matter in the wastewater. The N and P are readily taken up by the microorganisms and later released or are converted into soluble forms easily taken up by the growing grass. Also the maintenance of a relatively high air pore space throughout the root zone of the soil increases the oxygen transfer rate in this soil-plant system which further enhances the biological processes and lessens the possibility of anaerobic conditions arising. Parrlahti and Vartiovaara (1958) reported that in Finnish organic soils most of the important biological activities occurred in surface horizons where aerobic microorganisms, exchangeable nutrients, and living roots were concentrated.

The hemic material over sand had a much higher hydraulic conductivity than hemic material alone and thus very high rates of application were possible.

Experiments showed that the organic material should not be packed too dense. Bulk densities as high as 0.3 to 0.4 g/cc did not allow the water to percolate fast enough and thus limited the rate of application as well as the removal of P. Also there was too little air pore space for adequate biological activity. Air pore space less than 20% on the volume basis was too low for accelerated biological activity. Also in packing the organic soil material air pockets at the organic soil-sand interface and throughout the column had to be avoided. Air voids either as horizontal layers or pockets caused perching of water and reduced the effectiveness of the system.

Studies showed that the concentration of phosphate in the water applied was also a limiting factor. Using synthetic effluent at a P concentration of 40 to 50 mg/liter it was found that the adsorption effectiveness of the organic soils was greatly reduced. The trials using these high concentrations

Table 1. Equilibrium water content after saturation for hemic organic material.

Soil depth	Hemic material over sand*		Hemic material alone	
	Water content	Air pore space	Water content	Air pore space
cm	% on volume basis			
0 (surface)	55	31	72	14
5	55	31	74	12
10	55	30	76	10
15	56	30	78	8
20	56	30	80	6
25	57	29	82	4
30	57	29	86 (saturated)	0

* 30 cm hemic organic material over 60 cm fine sand; bulk density of organic material = 0. 2 g/cc or 12. 5 lb/cu. ft.

were ineffective after only 2 to 3 weeks of operation. When the concentration of P was reduced to 5 to 10 mg/liter the system worked very well. Later studies showed that the most effective system for removing P was one in which the carbon/phosphorus ratio was about 500:1 and the N/P ratio about 4:1. The research effort then concentrated on utilizing an average wastewater and determining the optimal application rate based on concentration of N and P in effluent and the uptake potential of specific grasses.

The data in Table 2 shows the removal of P and reduction of BOD using hemic organic material where wastewater was applied at 9 to 10 cm/day. After 12 weeks of operation the system showed very excellent phosphorus and BOD reductions. The P had been reduced to less than 0.05 mg/liter and BOD to less than 2.5 mg/liter. These very high removals are attributed to the high biological activity in these soils. Phosphorus is probably being removed partly by the soil, partly by microorganisms and also by the plants.

Table 2. Wastewater treatment using hemic organic material in a lysimeter.

Operation	Wastewater applied (cumulative)	PO_4 (P)		Biochemical oxygen demand (BOD)	
		Influent (secondary waste)*	Effluent (filtered)	Influent	Effluent
weeks	cm	mg/liter			
1	65	6. 5	0. 01	150	1. 0
2	130	4. 8	0. 03	-	-
3	195	6. 5	0. 01	145	2. 0
4	260	5. 1	0. 03	-	-
5	325	6. 8	0. 02	160	2. 5
6	390	8. 0	0. 05	-	-
7	455	6. 1	0. 05	145	1. 5
8	520	7. 9	0. 05	-	-
9	585	9. 5	0. 05	140	2. 0
10	650	8. 2	0. 05	-	-
11	715	6. 4	0. 05	-	-
12	780	6. 2	0. 05	135	2. 0

* Average application rate of secondary wastewater 9 to 10 cm/day.

Levin and Shapiro (1965) reported very high metabolic uptake of P by wastewater organsims which they attribute to high dissolved oxygen levels and high carbon content. Certainly one would expect this to occur in these low density organic materials over sand where aerobic conditions are always present and fresh carbon is being continuously added by applying wastewater. The accelerated microbial activity in these systems could easily explain the BOD reductions due to the rapid decomposition of added organic material in the effluent.

Farnham and Brown (1972) showed similar very effective removals of both phosphorus and BOD using sphagnum peat (fibric material) even after applying over 3,500 cm of effluent at a relatively high rate. Excellent reduction in pollutants was realized even though the grass crops did not grow as well on sphagnum peat as on the hemic organic materials. Sphagnum was too acid at first to establish the grass.

In order to maximize the uptake of N and P by grass crops it is suggested that the irrigation application rate be based on the concentration of those two elements in the wastewater used (Table 3) and the amount of these nutrients a particular crop may be expected to remove (Table 4). If, for example, the concentration of N and P in a wastewater was 15 mg/liter of N and 5 mg/liter of P and the grass used was quackgrass grown in a northern climate (growing season of 26 weeks) one could expect to remove 600 kg/ha of N and 150 kg/ha of P (Table 4). Extrapolating from Table 3, a weekly application of about 15 cm of wastewater would provide the grass with 650 kg

Table 3. Wastewater irrigation rates and nutrient additions with varying concentrations.

Wastewater applied	Concentration of nutrient in wastewater-seasonal additions (26 weeks)			
	2 mg/liter	5 mg/liter	10 mg/liter	15 mg/liter
cm/week	kg/ha			
5	25.4	63.1	126.2	189.3
10	50.8	126.2	252.4	378.6
20	101.6	252.4	504.8	757.2
30	152.4	378.6	757.2	1,135.8
40	203.2	504.8	1,009.6	1,514.4

Table 4. Nitrogen and phosphorus removal by grasses grown on organic soils with overhead irrigation of wastewater.

	Yield dry matter	Nitrogen	Phosphorus
	kg/ha	kg/ha	
Kentucky bluegrass (6 month season)	9,000	300	80
Quackgrass (6 month season)	18,000	600	150
Reed canary grass* (6 month season)	21,700	900	180

* Data from Henry et al., 1954.

of N and 189 kg P/ha during the 26-week growing season. The amount of N and P added about equals that removed by the crop and ratio of N to P (3:1) is desirable for most grasses. From the results of this and other studies (unpublished data) it is possible to increase the P uptake of grasses by increasing the amount of nitrogen applied. In one experiment, quackgrass grown on an organic soil removed 900 kg of N and 185 kg P/ha by applying wastewater at a rate of 60 cm/week. Data in Table 4 for reed canary grass grown on organic soils in Wisconsin as reported by Henry et al. (1954) showed that 900 kg of N and 180 kg P/ha was removed by reed canary grass after applying 175 cm of wastewater during the season. Over 21 metric tons/ha was the dry matter yield.

Much higher yields of grasses could be expected in warmer areas such as California, Arizona, and Florida where grasses would grow all year long under irrigation practices. With application of wastewater containing average N and P one could probably expect removals as high as 1,200 kg/ha of N and up to 300 kg/ha of P. Data from Florida by Burt and Snyder (1972) showed excellent yield increase by applying frequent applications of irrigation water containing very low concentrations of N and utilizing a "fertigation" technique on bermudagrass (*Cynodon dactylon* L. 'Tifgreen') (E. O. Burt and G. H. Snyder, 1972. Nitrogen fertilization of Bermudagrass turf through the irrigation system. Agron. Abstr., p. 61.).

SUMMARY AND CONCLUSION

This research has shown that under the proper conditions organic soils are very effective when used as wastewater treatment systems. They are particularly efficient in removing phosphorus and organic matter. The phosphorus removal process was attributed primarily to biological immobilization rather than to chemical adsorption by iron and aluminum. This was due probably to the high metabolic uptake of phosphorus by wastewater and soil organisms and subsequently the conversion of inorganic to organic forms of phosphorus. The phosphorus removal efficiency is excellent under moderate loading rates and over a few years operation time. It will probably decrease in efficiency with time and with increased loading rates. However, experiments have not been conducted long enough to determine the long run efficiency of the system. For this reason, it is suggested that grass crops be grown on the organic soils to assure the effectiveness of the system over a long period of time. Grass crops can regularly remove large amounts both nitrogen and phosphorus and they also function in maintaining a high infiltration capacity and help aerate the soils root zone.

Some advantages of organic soils for wastewater treatment are as follows:

1) An abundant supply of oxygen in the soil.
2) Very high rates of application possible.
3) Less possibility of surface clogging and reduced infiltration.

4) An excellent environment for microorganisms as there is ample food in wastewater.
5) A very good media for growing grass.
6) Makes possible the detention of nutrients for subsequent removal by plants and biological conversion to organic forms.

An organic soil wastewater treatment system utilizing suitable grasses is effective because it combines the best features of a soil system with that of a plant system for maximum nutrient removal. This system is particularly suited for use in small towns, campgrounds and lake recreation areas where more sophisticated and very expensive treatment systems are not feasible. The US Forest Service is presently using this system at two lake campgrounds in the Chippewa National Forest in Minnesota and are planning to use it in other areas soon.

LITERATURE CITED

American Public Health Ass. 1971. Standard methods for the examination of water and wastewater. 13th ed. American Public Health Ass., New York.

Bouwer, H. 1968. Returning wastes to the land: A new role for agriculture. J. Soil Water Conserv. 23:164-168.

Day, A. D., T. C. Tucker, and M. G. Vanich. 1962. Effect of city sewage effluent on the yield and quality of grain from barley, oats, and wheat. Agron. J. 54:133-135.

Farnham, R. S., and J. L. Brown. 1972. Advanced wastewater treatment using organic materials. Part 1. Use of peat and peat-sand filtration media. Int. Peat Congr., Proc. 4th. (Otaniemi, Finland) Vol. 4:271-286.

Henry, C. D. et al. 1954. Sewage effluent disposal through crop irrigation. Sewage Ind. Wastes 26:123-135.

Levin, C. V., and J. Shapiro. 1965. Metabolic uptake of phosphorus by wastewater organisms. J. Water Pollut. Contr. Fed. 37:800.

Paarlahti, K., and U. Vartiovaara. 1958. Observations concerning the microbial populations in virgin and drained bogs. Commun, Instit. Forestalia Fenn. 50:4, 38 p. [Finnish with Eng. Summary].

Parizek, R. R., L. T. Kardos, W. E. Sopper, E. A. Myers, D. E. Davis, M. A. Farrell, and J. B. Nesbitt. 1967. Wastewater renovation and conservation. Penn. State Studies 23, The Pennsylvania State University.

Rogstad, T., and W. Larson. 1969. The eutrophication-pollution situation in the Detroit lakes area. *In* Lake Eutrophication—Water Pollution Causes, effects and control. Minn. Water Resources Res. Center. Bull. 22:17-26.

Silvo, O. E. J. 1972. Same experiments on purification of wastewaters from slaughter houses with sphagnum peat. Int. Peat Congr., Proc. 4th. (Otaniemi, Finland) 4: 311-315.

Surakka, S., and A. Kamppi. 1971. [Infiltration of wastewater into peat soil.] Suo 22: 51-58. [Finnish-English Summary].

Taylor, A. W. 1967. Phosphorus and water pollution. J. Soil Water Conserv. 22:228-231.

Temple University. 1972. Green land, clean streams: The beneficial use of wastewater through land treatment. Center for the Study of Federalism, Temple University, Philadelphia, Pa.

Factors Affecting Forest Production on Organic Soils[1] 10

T. EWALD MAKI[2]

ABSTRACT

In major wood-producing countries throughout the world, organic soils occupy extensive areas within the 4 billion ha classed as predominantly forest land. With continuing decrease in land available for wood production, foresters have intensified their exploration of the possibilities of forest management on organic soils which in the existing "undisturbed" state are generally submarginal for growing usable wood on an economic scale. The pocosins or raised bogs of eastern North Carolina fall into this submarginal category, and the 31.6-thousand-ha Hofmann Forest in Jones and Onslow counties is representative of soil, vegetation, and hydrologic characteristics, as well as of forest management problems of pocosin ecosystems.

Experience encountered in tests of forest management practices on organic soils and results of some research on the Hofmann Forest are recorded in this paper. Special emphasis is given to drainage, water table behavior, pocosin vegetation, soil properties, and growth of pines in response to site preparation and amelioration. Changes in the pocosin environment as a result of forest management practices are postulated.

INTRODUCTION

Organic soils occupy extensive areas within the 4 billion ha (about 10 billion acres) classified as potentially productive forest land on this planet. Their occurrence, obviously, is more common in cool regions, but extensive accumulations can be found also at lower latitudes, even in subtropical climates.

Organic soils are very important in terms of wood production throughout the world. For example, in Finland the peatlands comprise 9.7 million ha, or 32% of the total land area; in Norway, 3.0; in Sweden, 7.8; in the USSR, 150; and in Canada, 112 million ha (Heikurainen, 1964). Stoeckeler (1961) estimated the organic soils (*Histosols*) of Alaska to extend over 40 million ha. In the Lake States, LeBarron and Neetzel (1942) have reported nearly 3.3 million ha of organic soils. Organic soils in eastern North Carolina and the southeastern corner of Virginia, comprise roughly 1 million ha. Essentially, in every locality where organic soils abound, excess water inter-

[1]Paper no. 4137 of the Journal Series of the North Carolina Agricultural Experiment Station, Raleigh.

[2]Schenck Professor of Forestry, Department of Forestry, School of Forest Resources, North Carolina State University, Raleigh.

feres with the growth of coniferous tree species and forest management operations.

Organic soils, moreover, are not the only wet soils that interfere with extensive development of forest resources. Along the Atlantic and Gulf coastal plains from southern New Jersey to east Texas, including the southern Mississippi River Valley, wetland forests occupy some 15 million ha (35 million acres). Within this vast area, moisture control is essential if woodland operations are to be feasible, and growth of conifers, as well as certain hardwoods, is to be optimum. Scattered within these wetlands are extensive areas of organic soils. For example, in eastern North Carolina alone, there are over 600 thousand ha (1.5 million acres) mainly in raised bogs (i.e., in bays and pocosins)[3]. Organic soils also predominate in such well-known wetland forests as the Dismal Swamp astride the Virginia-North Carolina boundary and the Okefenokee Swamp in southeastern Georgia. These two areas alone comprise nearly 1 half million ha (over 1 million acres).

Precise information on the extent and location of organic soils is available only for a few countries, mainly Denmark, Finland, Germany, Great Britain, Norway, and Sweden. In the USA, even excluding Alaska, mapping of organic soils under forest cover is incomplete. For example, in forests along sluggish streams in the Atlantic and Gulf Coastal plains bands of organic matter accumulations are found with concentrations and depths sufficient to qualify them as mucks (i.e., Saprists). However, many of these areas have not been mapped as such in existing surveys. As a first approximation for the 15 million ha of wetland forest of the southeastern USA, Histosols, with depths of at least 60 cm (2 feet) and organic matter concentrations of 30%, or higher, are estimated as occupying some four million ha.

Availability of land for wood production is decreasing in amount, and often also in quality, leading to intensified forestry interest in organic soils. Highways, rurban crawl and urban sprawl, reservoirs, golf courses, and similar developments are taking some of the best soil-sites out of forest production, with very small likelihood that any of this preempted land will return to commercial woodland use in the future. There is also a continuing loss of forest area as land is cleared for fields and pastures, usually representing the better remaining soils. This loss, however, is partly compensated through idling of impoverished land and its eventual colonization by forest growth. Even forest-covered organic soils are undergoing substantial reduction in acreage. As one example, an early estimate by Kearney (1901) suggests that the Dismal Dwamp occupied an area of 3,900 km^2 (about 1,500 square miles). Recently, Ramsey et al. (1970) estimated the total area of the Swamp to be only 1,700 km^2 (about 580 square miles). The wide disparity between these estimates may reflect some inaccuracies in surveys. Nevertheless, it is known that drainage, clearing for agriculture, and rurban encroachment have made heavy inroads at the periphery of this swamp within the past 70 years.

[3]*Bays* normally are elliptical-shaped upland bogs, usually circumscribed partly or entirely by sandy rims. The designation, *pocosin,* is often limited to the irregularly-shaped "swamp-on-hill" areas, usually interspersed by essentially parallel ridges and depressions.

The indicated shrinkage in the land base for forestry purposes has provided the impetus to foresters to intensify efforts toward improving forest production on organic soils, often found submarginal in their native state for the growing of tree species to commercial size. The organic soils of pocosins and bays of eastern North Carolina fall into this type of a submarginal category.

This paper deals mainly with forest management problems of Histosols in pocosins. It reports the results of studies conducted in the development and management of forest resources in the Hofmann Forest in Jones and Onslow counties, North Carolina, and in other places. In addition, it calls attention to problems of fire protection, prescribed burning, control of unwanted vegetation, and of trafficability in conducting forest management on organic soils which remain unfrozen year-round.

PROCEDURES

Aside from limited reference to data from studies conducted elsewhere, this paper reports data from research conducted on the Hofmann Forest. First, a brief description of the physical features of this forest may help to visualize some of the problems of forest management in a pocosin environment and some of the underlying reasons for the research conducted there.

The Hofmann Forest

This forest is basically a large pocosin, or raised bog, comprising 31.6 thousand ha (about 78,000 acres) at approximately 35° N latitude. It is "boxed in" by the Trent, the White Oak, and the New rivers. The raised character of the pocosin surface is etched by these river systems; tributaries of the Trent River flow northward, those of the White Oak River flow southeastward, and tributaries of the New River flow generally southwestward out of the Forest.

The Forest's most extensive area of very low productivity is the 4,000-ha (10,000-acre) Big Opening, which comprises roughly one-eighth of the total area, and supports only a sparse stand of stunted pond pine (*Pinus serotina* Michx.) and interspersed thickets of pocosin shrubs. The soils are entirely organic in character and range in depth from 100 cm (about 3 feet) to 200 cm. Organic soils predominate throughout the forest, with a comparable range in depth.

For many centuries severe wildfires have raged periodically through the forest, but precise records on fire occurrence are available only for the last four decades. A severe crown fire in April 1950 burned over 20,000 ha (about 50,000 acres). The latest large fire swept into the forest on the afternoon of 16 April 1972, and before it was brought under control on the following day, it had burned over 8.5 thousand ha (over 20,000 acres). It consumed much of the brush understory and the foliage on the pines over extensive areas, but left the blackened pine trunks standing. In spots throughout the forest, past fires have reduced accumulated organic matter as much as 30 to 60 cm.

The Hofmann Forest typifies forest production problems encountered on organic soils at lower latitudes. Long growing seasons and ample rainfall stimulate luxuriant growth of native shrub species, mostly of unwanted variety; this biomass complicates forest renewal, and it intensifies the hazard of wildfires and the difficulty of fire control. In the deeper organic soils, it is difficult and expensive to construct access roads to facilitate drainage or moisture control, and to carry out other land management operations, such as site preparation, fertilization, planting and seeding. Finally, there is need to remain alert to possible ecological impacts from drastic alteration of pocosin ecosystems.

Water-Table Measurement

Water-table fluctuations on the Hofmann Forest have been determined by means of replicated well lines installed at right angles to main drainage ditches or canals extending from them up to 427 m (1,400 feet). On each line at specified intervals (*vide infra* Table 3, column 1) well pipes of 5.1 cm inside diameter (ID), and perforated by 0.64-cm holes arranged in a spiral pattern throughout their length have been inserted vertically into the organic layer to a depth of 1.5 m. Elevations of the pipe tops and the surrounding ground level have been obtained with a Wye level for gaging trends in water-table gradients. Readings from the top of the pipes which protrude 5 to 7 cm above ground to the water surface in the pipe have been made periodically, and on occasions daily, to the nearest 0.025 cm.

Measurement of Tree Growth Response

Beginning in 1936, on variously prepared sites, pines and other coniferous species have been planted on the Hofmann Forest to observe tree growth response particularly to drainage. Except for the initial plantation, randomized complete block designs have been generally employed in these experimental plantings. Both survival and total height have been recorded at intervals, usually of several growing seasons, on the experimental plots until the trees have attained sufficient size to warrant measurement also of their diameter at breast height (dbh) measured 1.3 m (4½ feet) above ground level. Diameter and height data have been used with appropriate volume tables to determine volume growth response.

Soil Sampling

For major laboratory analyses, replicated soil samples have been taken at depths of 0 to 15 cm and 30 to 45 cm; these depths span the initial rhizosphere of most 1-year-old coniferous planting stock. A single replication has consisted of four subsamples of each depth zone, each extracted at about 2-m intervals within a given experimental plot. The subsamples have been dumped on a plastic sheet and mixed thoroughly to form the composite from which approximately 4 liters have been withdrawn for laboratory analyses. An abundance of roots and severe shrinkage on drying necessitate taking a relatively large volume of the forest-fresh muck into the laboratory.

Laboratory Analyses

The soil samples for which results are reported here have been analyzed for acidity (pH), available P and K, exchangeable Ca and Mg, organic matter, and cation exchange capacity. Acidity (pH) has been determined on a water sample using a glass electrode. Phosphorus was extracted with 0.05*N* HCl–0.025*N* H_2SO_4 solution, and the P content determined spectrophotometrically by the molybdovanadate method. Potassium was determined by flame photometer and Ca and Mg by atomic absorption spectrophotometry. Organic matter was determined by wet digestion, using 4.0*N* sodium dichromate and concentrated H_2SO_4.

RESULTS AND DISCUSSION

As background for results of studies on the Hofmann Forest a review of some research elsewhere may serve as a useful frame of reference and point to some of the complexities in management and utilization of Histosols.

Histosol Variation and Complexities

In the USA the classification of organic soils for forest production has not advanced very far. Considering that Histosols are derived from a common parent material, i.e., vegetation, the variation in character and composition is surprisingly large. It may be, of course, that the plant origin is the major basis for the observed variation and complexity in organic soils. Vegetation is influenced by climate, by climatic change over time, by the nature of the "substrate" over which it develops, by ground water inflow from the upland terrain surrounding the habitat, and by other factors related to these things.

Although comparisons of the formation of organic soils with development of mineral soils may be of doubtful validity, some speculation seems warranted. The important primary mineral groups are only three in number, *viz.*; the silica group, the feldspathic group, and the ferromagnesian group. In considering parent materials of organic soils, we might liken the primary mineral groups to primary vegetation groups, i.e., trees, shrubs, mosses, sedges, rushes, and grasses. Although these major groups of vegetation are not mutually exclusive, they predominate in various strata of peat deposits (*peat* is used here in a general, generic sense). This no doubt reflects differences in climate, degree of drainage, salinity, and other conditions that changed periodically during the time the organic materials accumulated.

In this context, it may be of interest to note some of the differences observed in three major categories of organic soils designated by Dachnowski-Stokes (1933) as (i) oligotrophic, (ii) mesotrophic, and (iii) eutrophic. The first category represents the most acid soils and also the poorest in nutrients. Some of the results of a study by Feustel & Byers (1930) are given in Table 1.

In this study (Table 1) the variable depth of sampling undoubtedly reflects the difficulty of securing suitable segments of the profiles for bulk density determinations. This difficulty is due to the presence of a surface mat of roots which make extraction of an undisturbed core impossible. There are major differences in acidity, in the concentration of nitrogen and calcium, and, in the surface layer, of aluminum. Differences in the concentration of ash after ignition are also apparent. However, the confounding effect of several factors including latitude, geologic history, occurrence and intensity of past wildfires, and other influences leave very little room for speculation about which part of the variation is caused by differences in vegetative origin of the organic surface.

Table 1. Chemical composition of peat from selected locations in the eastern United States (Adapted from Feustel & Byers, 1930).

Location and origin of sample	Approx. depth of sample	Bulk density	Acidity	Ash	N	P	K	Ca	Al
	cm	g/cc	pH	%					
Cherryfield, Maine	1-1.5	0.68	3.7	2.31	0.59	0.026	0.067	0.360	0.156
Oligotrophic (Sphagnum peat)	2-3	0.91	3.8	1.80	0.95	0.039	0.050	0.165	0.063
	3-5	1.10	3.6	1.55	0.82	0.035	0.025	0.136	0.092
Beaufort, N.C.	1-2	0.53	4.0	7.17	2.08	0.056	0.033	0.415	0.405
Mesotrophic (Ericaceous shrub peat)	2-4	0.85	4.0	2.33	1.71	0.030	0.025	0.172	0.123
	5-7	0.88	3.5	2.08	0.98	0.009	0.008	0.078	0.102
Belle Glade, Florida	0-1.5	--	5.3	8.54	4.24	0.096	0.058	2.12	0.079
Eutrophic (Sawgrass peat)	1.5-2.5	0.60	6.2	6.67	3.75	0.044	0.025	2.10	0.029
	11-13	0.97	6.3	10.67	3.72	0.017	0.033	3.72	0.117

Table 2. Nutrient concentration and acidity of the surface 20-cm layer of selected peatland types in Newfoundland (Adapted from Heikurainen, 1968).

Peatland type	Acidity	N	P	K	Ca	Mg
	pH			%		
Dwarf shrub bog	3.41	0.64	0.026	0.035	0.15	0.106
Sedge bog	3.89	1.54	0.046	0.027	0.07	0.043
Sphagnum fen	4.51	2.13	0.066	0.033	0.15	0.037
Brown-moss fen	5.16	2.01	0.092	0.054	0.24	0.121
Fen-like tree swamp	5.20	1.83	0.078	0.046	0.44	0.116
Herb-rich black spruce swamp	4.00	1.00	0.067	0.042	0.17	0.116
Range in peat types of Finland	3.5-5.5	1.2-2.4	0.025-0.100	0.015-0.070	0.25-1.35	0.03-0.15

A better comparison of the influence of vegetation types on surface layers of peat can be made on the basis of a more localized study (Heikurainen, 1968). Heikurainen described 11 vegetation types in peat bogs of Newfoundland based principally on existing dominant species from shrubs to trees. Some of his data given in Table 2 shows the range of nutrient concentration in the surface layer to a depth of 20 cm (about 8 inches).

A number of Heikurainen's samples were from raised bogs (internationally termed *ombrotrophic*) that receive most of their water from rain and snow. Others were from fens and other peatland types, (termed *minerotrophic*) in which outflow from surrounding uplands tends to make them richer in nutrients. His sampling was confined to the island portion of Newfoundland within a zone roughly between 48° and 50° north latitude. Here the climate is ameliorated enough by the presence of the ocean to rule out climatic influence as a major source of variation. Drainage and moisture economy and related geomorphic features, however, cannot be discounted because they influence the development of existing dominant vegetation communities which have supplied the parent material for at least 20 cm of the surface organic layer. Perhaps all that can be justifiably concluded in this work is that nutrient concentration appears to be associated with the type of vegetation occupying the bog surface, but cause and effect are difficult to assess.

Reference to the above studies have emphasized nutrient concentrations and variations among them. Some of the differences are large enough to affect tree growth (cf. Heikurainen, 1964; Tamm, 1964). However, in using organic soils for forest production, a number of other complexities must be considered. One of these is the variation in hydraulic conductivity (Boelter, 1965). Water storage characteristics of different peat types vary significantly, and the differences have important hydrologic implications (Boelter, 1964). In addition, from a practical standpoint in forest management operations, maintenance of trafficability under heavy loads, nominal control of the persistent burgeoning of brush and reed biomass, and difficulties encountered in mechanical site preparation and in applying prescribed burning as a silvicultural measure to minimize fuel build-up and to facilitate wildfire control are important considerations.

Organic soils at lower latitudes are as difficult to manage for forest production as any. The single item of trafficability is a good example. At higher latitudes where ground surface remains frozen for several months, the need for high standard access roads is minimized. In contrast, in southeastern USA access roads for utilization, protection, and other forest land management operations require a considerable capital outlay for construction and continued maintenance.

Examples of forest management on organic soils are available for study on the Hofmann Forest. This area is remote enough to have missed encroachment by agriculture for the past 250 years. Most of the problems encountered in attempts to develop a viable forest production enterprise on predominantly organic soils at lower latitudes are represented in the management of this forest. Not the least of the problems is the abundance of native pocosin shrubs.

The Hofmann Forest: Pocosin Vegetation

Even though excess water and periodic damage by wildfires have kept forest productivity at a low level, the capacity of the organic soils to generate biomass in the form of clumps and thickets of undergrowth shrubs is tremendous. Studies have shown that the shrub clumps, reeds, plus understory vegetation and litter in pocosins average 18,000 kg/ha or about 16,000 lb/acre (Wendel et al., 1962). The pocosin vegetation can be grouped into a dozen or more recognizable types that have significance in rating of fuel and blow-up fire potential. The principal species comprising these types include one or more of the following shrubs:

Swamp cyrilla or Titi	*Cyrilla racemiflora*
Fetterbush	*Lyonia lucida*
Common gallberry	*Ilex glabra*
Honeycup	*Zenobia pulverulenta*
Loblolly bay	*Gordonia lasianthus*
Red bay	*Persea borbonia*
Sweet pepperbush	*Clethra alnifolia*
Dangleberry	*Gaylussacia frondosa*
Leatherleaf	*Chamaedaphne calycaluta*
Lambkill	*Kalmia angustifolia*
Common greenbriar	*Smilax rotundifolia*
Bayberry	*Myrica cerifera*

Extensive reed "beds" are also present, often called switch cane (*Arundinaria tecta*). The viable overstory where it is present (and usually sparse) is dominantly pond pine.

The shrubs form dense clumps and thickets of highly flammable fuels with an abundance of waxy foliage that has the capacity to generate intense heat. Thus, wildfire control is extremely difficult, and prescribed burning is hazardous. Essentially all of the species sprout vigorously, so fire does not kill them even though the hottest fires frequently consume the biomass to groundline. The shrubs are also a major hindrance to disking and bedding in preparation for planting. Shrubs become minor in importance once a well-stocked stand of pine is established and has developed beyond the crown closure stage. At this stage of forest development only shade-tolerant shrubs survive and brush density is greatly diminished.

Pond Pine vs. Loblolly Pine

Pond pine has survived in the pocosin habitat primarily because of its capacity to sprout either from stumps or from dormant buds along the bole. However, its serotinous (closed) cones provide additional insurance. Seed within the closed pond pine cones is known to remain viable and of high germinative energy on the tree for 5 years or longer (Kaufman & Posey, 1953). Fires cause the serotinous cones to open, and an ample shower of seed is released on the freshly-burned surface wherever seed trees are abundant. Thus, in any month of the growing season, pond pine has the capability to restock a fresh burn from seed which has remained in storage, often for several years, in the closed cones.

When the bole of pond pine is killed by fire, or when the tree is cut, a large number of stump sprouts will form no matter what the size or age of the tree. However, as shown by Hafley,[4] a dominant sprout will emerge and develop into a mature tree only from stumps with groundline diameters less than 12 cm (about 5 inches).

If fire does not girdle the bole entirely, the dormant buds can regenerate a new crown, despite complete defoliation and even loss of twigs and larger branches. The remarkable growth recovery from fire damage has been demonstrated by Asher[5] who sampled radial growth of trees completely defoliated by the April 1950 Hofmann Forest fire. Asher's sample included trees ranging in age from 20 to 60 years and in dbh (diameter at breast height, i.e., measured 137 cm above general ground level) from 10 cm to 36 cm (about 4 to 14 inches). The 4-year periodic radial growth before the fire (1946–1949) averaged 0.985 cm in dbh and the growth from 1953–1956, after an initial recovery period (1950–1952) averaged 1.323 cm. The mean difference of 0.34 cm is significant at the 0.01 level suggesting complete recovery of the burned trees.

In current artificial afforestation or reforestation of pocosin sites, loblolly pine (*Pinus taeda* L.) and slash pine (*Pinus elliottii* Engelm. var.

[4]William L. Hafley, 1957. Some factors affecting initiation and growth of pond pine sprouts in a pocosin area of eastern North Carolina. M.S. thesis, North Carolina State University, Raleigh.

[5]William C. Asher, 1957. Some interrelationships of drainage, stand development, and growth of pond pine (*Pinus serotine* Michx.) on the Hofmann Forest. M.S. thesis, North Carolina State University, Raleigh.

elliotii) are usually used to replace the native pond pine, despite the nominal insurance pond pine provides against complete loss in the event of wildfire. On fertile upland sites both loblolly pine and slash pine are superior in growth rate and form of woods-run pond pine, but on organic soils both may require more fertilization, particularly with phosphorus, to sustain satisfactory growth rates. Genetically superior strains of pond pine are already being developed (B. J. Zobel, personal communication), but until they become operationally available, the species will be used very sparingly in forest renewal efforts on the raised bog sites. However, since there is no assurance that "blow-up" fires won't occur in the future, the prevailing practice of species substitution appears to be risky business.

Drainage or Moisture Control

Control of soil moisture for forest growth is essentially no different than that for other crops or plants. Drainage is necessary for obtaining best growth of all coniferous species mainly under two major and fairly distinct conditions: (i) in situations having high water tables, and (ii) in situations where excess surface water cannot percolate with reasonable rapidity below the major rhizosphere of trees as on level terrain where prolonged ponding frequently results.

With reference to water-table behavior in pocosins, drainage or moisture controls involves mainly the manipulation of high water tables. It is postulated, though not yet proved, that through most of the growing season the depth to water table is optimum for southern pines if it can be held about 45 to 60 cm below the soil surface. The low rate of hydraulic conductivity

Table 3. Average depth of water table in three soil types in the Hofmann Forest, North Carolina, during late winter and early spring before onset of major transpiration draft. Basis: 9 weekly measurements per mean.

Distance from drainage ditch	Area I Typic Umbraquult		Area II Typic Medisaprist		Area III "Shallow Medisaprist"‡	
	Canalside	Roadside†	Canalside	Roadside	Canalside	Roadside
m*	cm					
9.	79	20	66	22	46	48
18.3	69	21	61	22	46	48
36.6	61	19	55	22	39	47
54.9	56	19	53	22	37	48
91.5	51	15	51	22	37	48
128.0	46	13	50	22	38	48
207.3	41	11	50	20	39	48
280.5	39	9	49	14	36	48
Mean	55.2	15.9	54.4	20.8	39.8	48.0

* The fractional distance result from converting the original measurement from English to metric units; i.e., 30 feet is approximately 9.1 m, etc.

† The roadside observation wells have drainage blocked by the roadbeds in Areas I and and II, but not in Area III.

‡ Tentative designation; essentially a Portsmouth mucky loam.

in Saprists make maintenance of the water table, at a fixed level over any great distance from ditches or canals, difficult or impossible. However, with sufficient intensity of ditching coupled with control devices, and with manipulation of the microsite by bedding (i.e., drawing up elevated ridges by means of heavy, specialized disk plows), it seems possible to lower the water table and keep it at a prescribed level, while excess water is removed by runoff.

Some of the problems encountered in attempting adequate and reasonably uniform moisture control over extended areas are emphasized in a study by Gallup.[6] Considerable differences in water-table behavior are evident, depending on soil characteristics even in months when transpiration draft by vegetation is not high (Table 3). Obviously, not all soils respond equally well to the same intensity of drainage. Gallup's data show the typical arc-like pattern of the water table surface within a zone of about 300 m of a canal or ditch, but only where hydraulic conductivity is sufficiently high to promote lateral movement of water into the ditches. In Area I of Gallup's study, the soil is a Typic umbraquult (Buol, 1968)[7] with high concentration of organic matter in the surface 30 cm. The organic layer is underlain by clay loam which develops good structure where ditching lowers the water table by 50 cm or more. Area II consists mainly of Typic Medisaprist soil to depths of 150-200 cm. However, the entire Saprist profile is laced with root channels of previous forest stands forming numerous interstices through which water moves rapidly into ditches. In Area III, about 30 cm of the surface consists of Saprist material and the soil is tentatively designated a "shallow Medisaprist". Considerable sphagnum moss growth has developed on the surface. The mineral soil below the organic material is a very compact loamy fine sand grading to find sandy clay loam at about 140 cm, except on the roadside portion where the clay loam occurs in spots immediately below the organic horizon. Essentially no water moves through the mineral horizon. Hence drainage is impeded and takes place mainly through the organic horizon and the interface with the mineral matter. The sphagnum mat remains saturated throughout the year. Area III represents a situation that can be drained with difficulty, or not at all, by ordinary ditching practices.

The data in Table 3 also demonstrate the noticeable barrier effects of roads for Areas I and II. In Area III, the low hydraulic conductivity of the mineral horizon appears to make canal and road position immaterial. Even where the roadbed consists of saprist material with an abundance of roots it compacts so well that it serves effectively as a dam. If it were desired to lower water tables by pumping, such areas could be effectively dammed or "ringed" by roads, even where the organic matter exceeds several meters in depth.

---

[6]L. E. Gallup, 1954. Some interrelationships of drainage, water table, and soil on the Hofmann Forest in eastern North Carolina. M.S. thesis, North Carolina State University, Raleigh.

[7]Stanley W. Buol, 1968. Descriptions of soils in field tour notes for the 3rd North American Forest Soils Conference, 9-10 August 1968.

General Growth Improvement

In the Hofmann Forest, a sufficiently long record is not yet available to provide a meaningful assessment of improvement in usable wood production on the deeper organic soils. On soils with clay loam to clay subsoils, periodic measurements of loblolly pine and slash pine planted in 1936 have shown high production potentials within zones no greater than 80 m or so from drainage canals. Within these zones the best performance of loblolly pine after 27 years has been 430 m^3 of wood with bark, equivalent to about 16 m^3/ha per year (228 cubic feet/acre per year). Within a zone 100 to 180 m from the canal, in the same planting and time span, usable wood with bark has been 75 m^3 less/ha. This differential emphasizes the sensitivity of loblolly pine to adequacy of drainage, and calls to mind the water table behavior reported in Table 3.

Since marketable volumes have not yet accumulated on test materials on the deeper organic soils, results from them can be shown best in terms of tree height.

Effects on Growth from Drainage Alone

In the Big Opening of the Hofmann Forest, exploratory trials of bedding were undertaken in the mid-1950's to provide elevated microsites for test plantings of several species. Four-disc fire plows were used to draw up berms with a furrow between a given set of berms spaced about 3 m apart. Initial response of 1-year-old seedlings planted on berms was quite promising (Table 4). A meaningful comparison of seedling growth without any site preparation was not possible because of competition from pocosin shrubs, but assessment of seedlings planted on berms versus those planted at the furrow bottoms which were 45 cm nearer the existing water table was made. Two years after planting, the seedling heights on the berms compared favorably with heights attained on fairly fertile, well-drained upland mineral soils. However, 10 years after planting, the heights were no longer comparable, reflecting unfavorable site conditions including a low nutrient status of the soils.

Table 4. Height growth of loblolly, slash and pond pines 2 years versus 10 years after planting on berms and in furrows in Typic medisaprist soils of the Big Opening, Hofmann Forest, Onslow County, N. C.

Species	Two years after planting		Ten years after planting	
	On berm	In furrow	On berm	In furrow
	meters			
Loblolly pine	1.10	0.54	5.7	3.9
Slash pine	0.97	0.54	6.3	4.5
Pond pine	0.75	0.54	4.4	3.8

Table 5. Average fertility status and related information on undisturbed surface soil in the Big Opening of the Hofmann Forest, Onslow County, N. C. Basis: 4 composite samples per mean.

Fertility status, etc.	Sample depth	
	0-15 cm	30-45 cm
Acidity (pH)†	3.8	4.2
P (kg/ha)	30	6
K (kg/ha)	167	16
Ca (kg/ha)	175	75
Mg (kg/ha)	364	154
Organic matter (%)‡	25.9	28.1
Cation exchange capacity (meq./100 cm^3)	11.40	7.38
Volume weight (g/cm^3)	0.45	0.52

* All analyses were performed by the Soil Testing Division, North Carolina Department of Agriculture, using their standard research procedures.
† In water solution.
‡ By wet digestion method.

An indication of the character of the Big Opening surface soil and its low mineral status is given in Table 5. Surface samples (0–15 cm) of undisturbed brush-covered soil are compared with subsurface samples from a zone 30 to 45 cm beneath the same spots from which the surface samples were extracted.

The above data clearly indicate the low nutrient status of the highly acid pocosin soils. The concentration of organic matter as determined by the wet digestion method suggests a lower level of volatility for these soils than is actually the case. Loss on ignition at 600C for samples from depths down to 150 cm indicates that about 80 to 95% of the organic material does volatilize. Probably a very substantial portion of this loss is attributable to abundant charcoal and charred plant materials from fires occurring at frequent intervals during past centuries. Nevertheless, pocosin soils do have vast reserves of nitrogen. Total N in the surface 30 cm of the Big Opening ranges from about 0.7 to 1.0% suggesting an N reserve from 4,000 to 8,000 kg/ha in the "furrow-slice". Limited sampling has indicated this concentration of N to extend down to 150–200 cm, or essentially to the bottom of the organic material accumulation in soils of the Big Opening of the Hofmann Forest.

Mobilization of the N reserve is essential for attaining satisfactory tree growth, but other measures are also needed. Addition of P, Ca, possibly some other minerals, reduction in competition from shrub growth, and lowering of the water table are also major requirements for good growth of coniferous tree species.

Effect on Growth from Drainage, Site Preparation, and Fertilizing

Bedding, as defined earlier, is a form of site preparation that reduces competition, improves drainage, and elevates microsites on which both aeration and temperature are made more favorable for tree growth. Sites can also be improved by burning and disking to further reduce the effect of brush competition.

Burning and disking in combination with treatments of lime and phosphorus have been tested in the Big Opening. Lime was applied at a rate of 10,000 kg/ha of ground limestone, while phosphorus was applied at a rate of 392 kg/ha of P in the form of Calphos, (a colloidal phosphate containing over twice as much Ca as P). Nine years after planting the total heights of loblolly pine on soil subjected to burning and disking and the addition of P and Ca averaged 4.65 m, while heights on land similarly site-prepared but without the addition of minerals has attained an average of 1.7 m. This almost 3-m height differential is impressive only if compared with seedlings receiving no Ca and P. Compared with seedlings on better sites, where an average rate of 1 m in height per year is not unusual, even the best growth on treated sites is not as good as it might be. Bedding after addition of the Ca and P would have resulted, undoubtedly, in improved height growth of loblolly in this pocosin site.

Basic differences in native soils can make a large difference in growth performance of planted pine. As an example, a site where the soil was a Terric medisaprist with only about 60 cm of organic matter underlain by loam to clay loam was adequately drained, chopped, burned and bedded before planting of loblolly pine. Three years after planting the pines averaged 3 m in total height in a zone within 10 m of the drainage ditch, but 2.3 m in a zone 30 m beyond the ditch. Although these data again emphasize the sensitivity of loblolly pine to excessive soil moisture the growth even at 30 m from the ditch still is considerably better than the 1.8 m in total height attained by this species 3 years after planting on average Piedmont upland sites. However, assessment of total heights 7 years after planting indicated that the very rapid early growth rates have not been sustained. The average annual rate decreased to slightly less than 1 m/year, although normally an increased rate of height growth is expected following the initial 2- to 3-year period after planting. Addition of P and Ca might stimulate height growth but probably will not restore it to the rate exhibited in the first 3 years after planting. Exploratory tests are now underway to determine whether incorporation of Ca and P at the time of bedding will result in sustained high rate of growth.

Changes in Pocosins Resulting from Forest Management Practices

In attempts to improve forest production on organic soils, one of the more persistent questions concerns possible damage to the "natural" pocosin ecosystem from disturbances entailed in the application of forest management practices. Development of artificial drainage clearly alters pocosins rapidly and drastically since it is a prelude to chopping, burning, bedding, or related site preparation measures. These measures bring about changes in terrain and soil characteristics many times more rapidly than would take place if natural processes were allowed to proceed at the usual rate.

Invariably, the development of artificial drainage in a pocosin or any other wetland tract is accompanied by road and trail construction making the area more accessible to a wider spectrum of human use. Such use could increase the incidence of wildfires and facilitate poaching of deer and bear.

Modification of pocosins can have some trade-offs assumed to be beneficial. Site and drainage manipulations are normally associated with noticeable changes in vegetative cover. Relatively unpalatable thickets of titi, lambkill, fetterbush and the like give way to such plants as broomsedge (*Andropogon virginicus*), blackberry (*Rubus* spp.), and blueberry (*Vaccinium* spp.), and ultimately to tall pines with abundant crops of mast. Experience on the Hofmann Forest suggests that in this altered environment there are more deer, quail, rabbits, and songbirds than were observed when the land was occupied by the "native" vegetation in its "undisturbed" state. The observed changes appear desirable in that they seem to result in a more diversified productivity than is apparent in a system where wildfires periodically destroy the vegetation and modify the soil.

Questions also persist as to whether drainage increases the rate of water removal, thus reducing ground-water recharge. The stratigraphy of pocosins would appear to rule out any appreciable ground-water recharge because the existing organic matter accumulations are underlain by a compact, indurated layer of mineral material. The layers of structureless clay beneath the organic and indurated mineral layers would appear to permit little, if any, water to percolate downward. If downward drainage had proceeded at a reasonable rate, pocosins would not exist since the surface wetness would not suffice and persist long enough to permit accumulation of peat.[8] These considerations make it appear that ground-water recharge is not jeopardized by current drainage measures nor by associated site preparation practices.

Questions raised as to accelerated eutrophication as a consequence of fertilization combined with drainage have not been resolved. The high exchange capacity of organic soils coupled with low hydraulic conductivity suggests that transport of nutrients from pocosin ecosystems is very slow and probably cannot be accelerated much above natural, background rates.

Finally, the question of subsidence is usually raised when use of organic soils is discussed. Subsidence associated with farming on organic soils has proved troublesome in many instances, and has often progressed at spectacular rates. For example, in the Sacramento-San Joaquin delta region, soils have "disappeared" at the rate of nearly 8 cm/year over a 28-year period (Weir, 1950). In Finland, Lukkala (1951) reported that 35 years after drainage of a forest swamp, compression of the organic layer (of fibrist material) varied between 21 to 41 cm. In most instances in forest production operations the compression of this order would appear to result mainly from break-

[8] J. Lamar Teate, 1968. Some effects of environmental modification on vegetation and tree growth in a North Carolina pocosin. Ph.D. dissertation, North Carolina State University, Raleigh.

down and deterioration of the surface raw root mat. This is a desirable outcome. Observations and repeated measurement of ground elevations on well lines in different soil types on the Hofmann Forest that have been drained for 30 years or longer have provided no indications that rapid subsidence will ensue from drainage or from the associated site preparation measures discussed.

SUMMARY AND CONCLUSIONS

Organic soils occur in many of the major wood-producing countries of the world. The total area amounts to several hundred million hectares. In the USA, land area available for forest production continues to decrease in amount. Thus, foresters are becoming more interested in organic soils. Large areas of such soils have not yet been taken up by agricultural enterprises. The pocosins or raised bogs of eastern North Carolina and elsewhere are in this available category, but in their native, "undisturbed" state, they are generally submarginal for growth of marketable wood. With protection from wildfires, adequate drainage, fertilization, and related silvicultural measures, organic soils offer considerable promise for development of viable forest production enterprises. However, the problems faced are as formidable as those that have kept these areas relatively free of agricultural encroachment.

Some of the problems confronting forest production on pocosin sites are discussed in a framework of a third century of study and experiences on the Hofmann Forest, a 31,000+ ha area of predominantly organic soils in eastern North Carolina. Major points emphasized are:

1) The pocosin vegetation is a formidable adversary comprising extremely flammable shrub communities with biomass and associated litter averaging some 18,000 kg/ha, oven-dry. The shrub mass is capable of vigorous sprouting, interferes with disking and bedding in advance of planting, and offers intense competition to tree seedlings.
2) The possibilities for developing desired control of water table levels are minimized by the low hydraulic conductivity of Saprists which predominate in bays and pocosins.
3) Growth of pine species is considerably improved by lowering water tables to 45 to 50 cm below the surface during the growing season. However, because of the limited distance to which a given ditch can lower the free water surface, optimum conditions for pine growth can be created on only a rather narrow zone (100 m, or so) paralleling the drainage channel.
4) Shallow (up to about 60 cm) organic soils underlain by loams and clay loams, or equivalent textures, are usually excellent sites for growth of pine species if the sites are adequately drained and prepared by chopper leveling of the fuel mass followed by broadcast burning, bedding and other measures as needed prior to planting.

Subsequent fertilization may be necessary to sustain the indicated initial rapid early growth rate.

5) Deeper organic soils generally do not sustain growth of pines at acceptably rapid levels. These soils are extremely acid and universally low in certain nutritive elements especially phosphorus. Bedding after addition of calcium and phosphorus prior to planting gives promise of acceptable growth of pines, but the optimum combinations of treatments to mobilize the large reserves of nitrogen and to control the water level have not yet been developed.
6) Drainage and associated site preparation measures appear to bring about what are usually considered to be desirable changes in the pocosin ecosystems. The systems seem to produce greater amounts of food species that deer, quail, rabbits, song birds, and related wildlife appear to prefer.
7) Because of the nature of pocosin stratigraphy, development of artificial drainage, plus site preparation, is not likely to affect ground-water recharge appreciably.
8) Subsidence of the organic profiles under forest production operations appears minimal and clearly does not entail the same hazards that face farming pursuits on Histosols.

ACKNOWLEDGMENTS

The author is indebted to former graduate research assistants Wm. C. Asher, L. Edward Gallup, Wm. L. Hafley, and J. Lamar Teate for use of data from their dissertation researches, to J. G. Hofmann, James Huff, W. W. Wicks, and other staff members of the Albermarle Paper Mfg. Corp. (now a Division of Hoerner-Waldorf) for the many experimental field installations over the past 22 years, and to the Soil Testing Division of North Carolina Department of Agriculture for soil analyses.

LITERATURE CITED

Boelter, D. H. 1964. Water storage characteristics of several peats *in situ*. Soil Sci. Soc. Amer. Proc. 28:433–435.

Boelter, D. H. 1965. Hydraulic conductivity of peats. Soil Sci. 100:277–230.

Dachnowski-Stokes, A. P. 1933. Peat deposits in the U.S.A. Handbuch der Moorkunde 7:1–140.

Fuestel, Irvin C., and Horace G. Byers. 1930. The physical and chemical characteristics of certain American peat profiles. USDA Tech. Bull. No. 214. 27 p.

Heikurainen, Leo. 1964. Improvement of forest growth on poorly drained peat soils. p. 39–113. *In* John A. Romberger & Peitsa Mikola (ed.) International review of forestry research, Vol. 1. Academic Press, New York.

Heikurainen, Leo. 1968. Peatlands of Newfoundland and possibilities of utilizing them in forestry. Forest Research Laboratory, St. John's Newfoundland Information Report No. N-X-16, 56 p.

Kaufman, Clemens, M., and Henry G. Posey. 1953. Production and quality of pond pine seed in a pocosin area of North Carolina. J. Forestry 51:280–282.

Kearney, T. H. 1901. Report on a botanical survey of the Dismal Swamp region. USDA, Div. of Botany Rep. No. 5, p. 321.

LeBarron, R. K., and J. R. Neetzel. 1942. Drainage of forest swamps. Ecology 23: 457–465.

Lukkala, O. J. 1951. Kokemuksia Jaakkoinsuo koeojitus-alueelta. (Experiences from Jaakkoin swamp experimental drainage area). Comm. Inst. For. Fenn. 39:1–53.

Ramsey, Elmer W., Kenneth R. Hinkle, and Laurence E. Benander. 1970. Waters of the Dismal Swamp. Va. J. Sci. 21:81–83.

Stoeckeler, J. H. 1961. Soil and water management for increased forest and range production. Soil Sci. Soc. Amer. Proc. 25:446–451.

Tamm, Carl Olof. 1964. Determination of nutrient requirements of forest stands. p. 115–170. *In* John A. Romberger and Peitsa Mikola (ed.) International review of forestry research, Vol. 1. Academic Press, New York.

Weir, Walter W. 1950. Subsidence of peat lands of the Sacramento-San Joaquin Delta, California. Hilgardia 20(3):37–56.

Wendel, G. W., T. G. Storey, and G. M. Byram. 1962. Forest fuels on organic and associated soils in the Coastal Plain of North Carolina. US Forest Service, Southeastern Forest Experiment Station Paper No. 144, 46 p.